Condensed Silica Fume in Concrete

Authors

V. M. Malhotra, B.Sc., B.E., D.D.L.(Hon.)
Head, Construction Materials Section
CANMET/EMR
Ottawa, Ontario, Canada

V. S. Ramachandran, Ph.D., D.Sc.
Head, Materials Section
National Research Council
Ottawa, Ontario, Canada

R. F. Feldman, M.A.Sc., D.Sc.
Building Materials Section
Institute for Research in Construction
National Research Council
Ottawa, Ontario, Canada

Pierre-Claude Aïtcin, Ph.D.
Professor
Department of Civil Engineering
Faculty of Applied Science
University of Sherbrooke
Sherbrooke, Québec, Canada

877955

Mitchell Memorial Library
Mississippi State University

CRC Press, Inc.
Boca Raton, Florida

Library of Congress Cataloging-in-Publication Data

Condensed silica fume in concrete.

  Includes bibliographies and index.
  1. Concrete--Additives. 2. Silica fume.
I. Malhotra, V. M.
TP884.A3C67   1987   666′.893   87-11635
ISBN 0-8493-5657-1

Direct all inquiries to CRC Press, Inc., 2000 Corporate Blvd., N.W., Boca Raton, Florida, 33431.

© 1987 by CRC Press, Inc.

International Standard Book Number 0-8493-5657-1

Library of Congress Card Number 87-11635
Printed in the United States

PREFACE

It is estimated that 80% of the concrete produced in North America contains one or more types of chemical and/or mineral admixtures. Admixtures confer beneficial effects to concrete, including enhanced frost and sulfate resistance, controlled setting and hardening, improved workability, increased strength, etc. Mineral admixtures are finely divided materials that are added to concrete in relatively larger amounts by weight of cement. Although raw or calcined natural minerals (pozzolans) are used in some parts of the world, many industrial by-products are rapidly becoming the primary source of mineral admixtures in use today.

Volatilized silica known as silica fume or condensed silica fume (SCF), a by-product in the production of metallic silicon or ferrosilicon, because of its high reactivity potential (surface area 20 to 23 m^2/g), has attained prominence in recent years. Condensed silica fume not only acts as an efficient pozzolan by reacting with the lime liberated during the hydration of Portland cement, but also influences the rate of hydration and possibly also reacts with the hydrated silicate product.

The first investigations on condensed silica fume began in the late 1940s and since then Norway has made significant advances on various aspects of silica fume as follows:

1948—51: The first tests were made on the use of CSF by Markestad at the Norwegian Technical University. The first technical note was published by Bernhard in 1952 in the Norwegian journal *Betong Idag*.

1968: Markstad's paper "Addition of Silica Dust to Concrete" at the Nordic Concrete Research Congress, Gothenburg.

1969: Testing of CSF at Concrete Laboratories, Norcem's Dalen Cement Plant.

1971: Joint work by Elkem A/S and A/S Norcem. Discussion on the use of CSF in Publ. No. 86 (Norwegian Geotechnical Institute).

1972: Internal Norcem Report by Bratlia on "Use of CSF in Ready Mix Concrete". Establishment of Project Group by Norcem.

1973: Norcem Reports on the Use of CSF in Prestressed Roof and Floor Units, compacted CSF in ready-mix concrete, etc.

1975: Norcem Reports: CSF in cement screws, CSF pellets, etc.

1976: Development of Corrocem, a product containing CSF as a key ingredient. Trondheim Mortelverk signed agreement with Ida Lilleby for making CSF.

1978: Norwegian standards 3420a and 3474 were amended to allow the use of CSF (8% maximum).

1979: Norcem started a separate organization, "Concrete Technology Department".

1980: Establishment of Elkem Chemicals.

Contributions made by other countries were relatively minor, at least in the period before the 1970s:

1971: A Japanese patent application dealt with the use of CSF to improve impermeability in concrete.

1973: Fiskaa Works in Norway became the first silica fume producers.

1975: A paper entitled "Possible Uses of CSF" was presented by Airea of Sweden at the ferrosilicon conference in Houston.

1976: A competition for the use of CSF by the European Ferrosilicum Purchasers was organized. The first commercial production of CSF was accomplished in Sweden.

Research organizations in Canada and the U.S. initiated work on CSF in 1980. The leading organizations in Canada are CANMET, University of Sherbrooke, and Division of Building Research, National Research Council. In the U.S. the Waterways Experiment Station is one of the leading organizations involved in CSF. The first major international conference on the use of silica fume in concrete was sponsored by CANMET and the American Concrete Institute in 1983 in Montebello, Canada and a Second International Conference was held in Madrid, in April 1986.

This book presents the present state of knowledge on the production, properties, and use of CSF in concrete and is divided into 12 chapters.

The first three chapters cover the production, types of silica fume, the physical characteristics and product variation, and the handling aspects of CSF. The material in the chapters serves as a basis for the practical use of CSF.

Chapters 4 and 5 cover the most important aspects of the hydration reactions in cement and the physical and mechanical properties of cement pastes. They serve as a background to explain the processes taking place in concrete containing CSF.

Chapter 6 discusses the physical, chemical, and durability properties of mortars containing CSF and explains these based on the properties of the pastes.

The next three chapters deal with the properties of fresh concrete, properties of hardened concrete, and durability of CSF concrete. The role of different dosages of CSF in concrete is emphasized.

The applications and standards form Chapters 10 and 11. The importance of CSF in specialized applications is explained.

The book concludes with a chapter on the biological and health effects concerning the use of CSF.

Each chapter contains a separate list of references. Full reference to the publications should permit the reader to delineate the contents. A list of additional references is given at the end for further reading. They are not referenced under any of the chapters.

V.M.M.
V.S.R.
R.F.F.
P.C.A.

THE AUTHORS

V. M. Malhotra, B.Sc., B.E., D.D.L.(Hon.), is Head, Construction Materials Section, Canada Centre for Minerals and Energy Technology (CANMET), Energy, Mines, and Resources Canada, Ottawa, Ontario. He has been with CANMET for 25 years.

Dr. Malhotra is an honorary member of the American Concrete Institute (Detroit) and the Concrete Society (London). He is also Honorary Fellow of the Institute of Concrete Technology (London), fellow of the Canadian Society for Civil Engineering, and fellow of the Engineering Institute of Canada. He is the recipient of a number of awards from the American Concrete Institute and ASTM for his contributions in the area of concrete technology. He is a member of numerous technical and administrative committees of the American Concrete Institute, ASTM, the Canadian Standards Association, and the International Standards Organization.

He is author/co-author of several books and is editor of numerous special publications of the American Concrete Institute. His current research interests include the role of supplementary cementing materials in concrete technology, concrete for offshore structures, and non-destructive testing of concrete and concrete durability.

V. S. Ramachandran, B.Sc. (1949), M.Sc. (1951), D.Phil. (1956), D.Sc. (1981), C. Chem. (Lond.), F.R.S.C. (Lond.), F.I. Ceram. (Lond.), F.A.C.S. (U.S.A.), was born in Bangalore, India. He was a Senior Scientific Officer at the Central Building Research Institute, Roorkee, India, until 1968 when he became a Research Officer in the Division of Building Research, National Research Council Canada. He is presently the Head of the Building Materials Section.

He has been engaged in research for 33 years and has made significant contributions in catalysis, clay mineralogy, lime, gypsum, and cement chemistry and concrete technology. He has published five books (two translated into Russian and one into Chinese): *Differential Thermal Analysis in Building Science, Application of DTA in Cement Chemistry, Calcium Chloride in Concrete, Concrete Science,* and *Concrete Admixtures Handbook.* He has also contributed 13 chapters to other books. He has published about 130 research papers.

Dr. Ramachandran is a Fellow of the Royal Society of Chemistry, Ceramic Society, U.K., and the American Ceramic Society. He is a member of the American Society for Testing and Materials, the American Concrete Institute, and the International Confederation of Thermal Analysis. He has been actively engaged as the chairman of various committees of the American Ceramic Society and is a member of two Steering Committees of the Canadian Standards Association. He has been a member of the Board of Abstractors of the American Ceramic Society for 16 years, and that of the Chemical Abstracts Service for 13 years. He has been a chapter contributor to the annual publication of the American Ceramic Society, *Cements Research Progress,* for 10 years. He is a member of the Editorial Board of the *Journal of Materials and Structures,* France. He is a full member of the Advisory Committee of RILEM. He is a member of the Concrete Research Council of the American Concrete Institute and has been a member of the organizing committees of several International Conferences.

Dr. Ramachandran is a recipient of a plaque from the Editorial Board of the journal, *Il Cemento,* Italy, for his contribution, a nominee for the Mettler Award, instituted by the International Confederation of Thermal Analysis (U.S.), and received an award for outstanding work in concrete technology from the University of Nuevo Leon, Mexico.

R. F. Feldman is Senior Research Officer in the Building Materials Section, Institute for Research in Construction, National Research Council Canada, Ottawa, Ontario.

Dr. Feldman received his B.Sc. (Chem. Maths. Physics) and the diploma in Chemical Technology from the University College of West Indies, London, 1956-57, and was awarded the M.A.Sc. from the University of Toronto in 1959 and the D.Sc. from the University of London in 1974.

Dr. Feldman's professional memberships include Registered Member of the Professional Engineers of Ontario (since 1960); the CSA Committee A-23 on Concrete Materials and Methods of Concrete Construction and the ASTM Committee C-1 on Portland Cement; Fellow of the American Ceramic Society, 1984 Chairman of the American Ceramic Society, Cements Division; and he represents Canada on three RILEM committees.

He is a past recipient of the Plummer gold medal for Best Paper in *Engineering Journal*. His literary achievements include presentation of the principal paper at the 7th International Symposium on Chemistry of Cement in Paris, 1980; the keynote address at the 8th International Congress on the Chemistry of Cement in Rio de Janiero, 1986; he has authored over 100 publications in scientific journals and co-authored one other book, *Concrete Science* (translated into Russian and Chinese). Dr. Feldman recently toured China on the invitation of the Chinese government as world expert to present 16 lectures and discussions on concrete durability and cement research.

Pierre-Claude Aïtcin is Professor of Civil Engineering at the Université de Sherbrooke, Sherbrooke, Québec, Canada.

He obtained his Ph.D. from the Université de Toulouse, France, in 1965.

He is currently a member of ACI Committees 226 and 363, ASTM Committees CO1 and CO9 and of CSA Subcommittee A-231 on Condensed Silica Fume.

Professor Aïtcin has authored or co-authored over 107 publications in the field of concrete. He is also the author or co-author of nine books on concrete technology.

His main fields of interest in concrete technology are high compressive strength concrete, the use of industrial by-products in concrete, and concreting practices in the Arctic.

ACKNOWLEDGMENTS

We would like to thank Dr. Jacques Dunnigan for the contribution of Chapter 12, Ms. Vasanthy Sivasundaram for the compilation of the list of extended bibliography and proofreading parts of the manuscript, Mr. Frank Crupi of the Division of Building Research, National Research Council Canada for providing drawings for Chapters 4 through 6, and Mr. Ray Chevrier of CANMET, Energy, Mines and Resources Canada for drafting work associated with Chapters 7 through 10.

ABBREVIATIONS

Cement nomenclature:

$C = CaO$, $S = SiO_2$, $H = H_2O$, e.g., $C/S = CaO/SiO_2$ ratio; $CH = Ca(OH)_2$, C-S-H = X $CaO \cdot Y\ SiO_2 \cdot ZH_2O$

W/(C + SF)	=	water cement + silica fume
W/S	=	water/solid ratio
W/C	=	water/cement ratio
IR	=	infrared analysis
SEM	=	scanning electron microscopy
XRD	=	X-ray diffraction
LOI	=	loss on ignition
CSF	=	condensed silica fume

Condensed silica fume and silica fume are used interchangeably

TABLE OF CONTENTS

Chapter 1

PRODUCTION AND TYPES OF CONDENSED SILICA FUME

I. INTRODUCTION

Condensed silica fume is a by-product of the manufacture of silicon or of various silicon alloys which are produced in so-called "submerged-arc electric furnaces".

A series of complex chemical reactions occur in the furnace as shown in Figure 1. Condensed silica fume particles appear to be formed by the oxidation and condensation of the gaseous silicon suboxide, SiO, which is formed in the reaction zone. When solidified condensed silica fume particles are entrained by the reaction gas or the reaction gas-air mixture coming from the furnace, they are collected in a dedusting system, generally in fabric bag filters.

As in many other highly technological industries, the silicon alloy producers have developed a whole range of specific products for specialized applications. Moreover, the customer's specifications for silicon alloys have become more demanding so that a given producer must use different raw materials to produce a certain type of alloy. As different types of silicon alloys have been produced in submerged electric arc furnace, there have been correspondingly different types of condensed silica fume, just as there have been different types of fly ashes and slags, depending on the type of coal burned or the type of iron ore processed.

II. PRODUCTION OF CONDENSED SILICA FUME

The type of alloy produced in a given furnace greatly influences the chemical composition of the condensed silica fume recuperated in the bag-house because the temperature and the chemical reactions in the furnace depend on the type and amount of metal alloyed to silicon and on the impurities present in the metal or the ores introduced in the burden.

The use of wood chips in the burden can also influence to some extent the chemical composition of condensed silica fume, especially its carbon content, and also the loss on ignition (LOI) and the alkali content.

The design of the furnace, with or without a heat recovery system, not only influences the color of condensed silica fume, but also its chemical composition, especially its carbon content (Figure 2). When the furnace is equipped with a heat recovery system gases leave the top of the furnace at about 800°C so that most of the carbon is completely burned. In a conventional furnace, gases leave the furnace at about 200°C so that unburned carbon particles and some wood chips will be collected in the bag-house with condensed silica fume.

In addition, the chemical composition of quartz and coal, the two major components of the burden of the submerged-electric arc furnace, influences to some extent the chemical composition of condensed silica fume. Although these materials are generally of high purity, they have to be very carefully selected and controlled.

Generally as the amount of silicon increases in the final product the level of chemical impurities in condensed silica fume decreases.

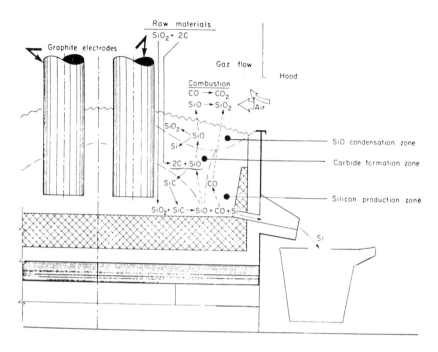

FIGURE 1.   Chemical reactions taking place in the reaction zone of a furnace.

III. DIFFERENT TYPES OF SILICON ALLOYS

Silicon alloys must comply to different ASTM standards as follows:

ASTM A100-80	Standard Specification for Ferrosilicon
ASTM A463-64 (1980)	Standard Specification for Silicon-manganese
ASTM A495-76 (1983)	Standard Specification for Calcium-Silicon and Calcium-Manganese-Silicon

The main types of silicon alloys presently produced in submerged-arc electric furnaces are

1. Silicon and ferrosilicon of various Si contents.
    *   Silicon metal; it is a metal grade containing more than 96% of silicon
    *   Ferrosilicon 90% containing from 92 to 95% of silicon
    *   Ferrosilicon 85% containing from 83 to 88% of silicon
    *   Ferrosilicon 75% containing from 74 to 79% of silicon
    This type is probably the most widely produced alloy
    *   Ferrosilicon 65% containing from 65 to 70% of silicon
    *   Ferrosilicon 50% containing from 47 to 51% of silicon
    This type of alloy is almost exclusively produced in the U.S.
    *   Ferrosilicon 22% containing from 22 to 24% of silicon
    *   Ferrosilicon 15% containing from 14 to 17% of silicon
2. Ferrochrome-silicon
    *   FeCrSi—48% containing 46 to 49% of silicon
    *   FeCrSi—43% containing 41 to 45% of silicon
    *   FeCrSi—40% containing 38 to 42% of silicon

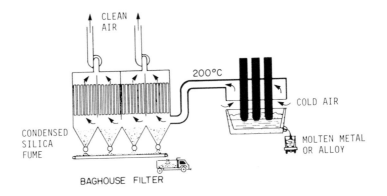

WITHOUT A HEAT RECOVERY SYSTEM

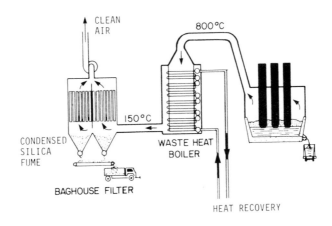

WITH A HEAT RECOVERY SYSTEM

FIGURE 2.   Schematic representation of silicon plants with and without a heat recovery system.

3.   Calcium-silicon containing from 60 to 65% silicon
   • CaSi alloy is produced through the reduction of a mixture of quick lime and silica by coal
4.   Calcium-manganese-silicon
   • CaMnSi containing 53 to 59% of silicon
5.   Siliconmanganese
   • SiMn containing 18.5 to 21, 16 to 18.5, or 12.5 to 16% silicon
6.   Magnesium ferrosilicon
   • FeMgSi—9% magnesium with 44 to 48% silicon
   • FeMgSi—5% magnesium with 44 to 48% silicon

Other types of silicon alloys are also produced but in very small amounts, or occasionally, by some silicon alloy producers so that the corresponding amount of condensed silica fume produced is not significant. Not all these different silicon alloys are produced to the same extent. Table 1 shows the estimates of the worldwide production in 1982.

The amount of condensed silica fume produced from 1 t of silicon or silicon alloy

Table 1
ESTIMATES OF WORLD PRODUCTION OF SILICON ALLOYS IN 1982[1]

	Silicon metal	FeSi-90%	FeSi-75%	FeSi-50%	FeCrSi	CaSi	SiMn
t/year	436,940	←	2,684,700	→	NA	NA	NA

Table 2
CHEMICAL COMPOSITION OF CONDENSED SILICA FUMES[3]

Chemical composition (wt%)

Component	Si	FeSi-75%	FeSi-75% (heat recovery)	FeSi-50%	FeCrSi	CaSi	SiMn
SiO_2	94	89	90	83	83	53.7	25
Fe_2O_3	0.03	0.6	2.9	2.5	1.0	0.7	1.8
Al_2O_3	0.06	0.4	1.0	2.5	2.5	0.9	2.5
CaO	0.5	0.2	0.1	0.8	0.8	23.2	4.0
MgO	1.1	1.7	0.2	3.0	7.0	3.3	2.7
Na_2O	0.04	0.2	0.9	0.3	1.0	0.6	2.0
K_2O	0.05	1.2	1.3	2.0	1.8	2.4	8.5
C	1.0	1.4	0.6	1.8	1.6	3.4	2.5
S	0.2		0.1				2.5
MnO		0.06		0.2	0.2		36.0
LOI	2.5	2.7		3.6	2.2	7.9	10.0

varies according to the type of alloy. For example, the production of 1 t of silicon generates about 0.60 t of condensed silica fume whereas 1 t of FeSi-75% alloy generates from 0.20 to 0.45 t.[2]

IV. CHEMICAL COMPOSITION OF CONDENSED SILICA FUMES

In Table 2 the average chemical composition of the most widely produced condensed silica fume is presented. The percentage of Si in various silica fumes shows a wide variation. The silica fume from the manufacture of SiMn alloy has the lowest SiO_2 content (25%), whereas that from Si production has the highest content (94%).

V. BLENDS

Frequently, more than one type of silicon alloy is produced at the same time in a given plant, and different types of condensed silica fumes are simultaneously collected in the bag-house.

As condensed silica fume has been, until recently, considered as a by-product without any commercial value, or even as a waste product with a high disposal cost, the dust removal system of silicon and silicon alloy plants has not always been designed to separately collect different types of condensed silica fumes produced at the same time by the different furnaces. For this reason some condensed silica fumes presently collected in some plants can be a mixture of different types. Depending on the type of condensed silica fumes collected and their respective proportions, some types can or cannot be used in cement and concrete. For example, the use of a mixture of Si and

FeSi-75% condensed silica fume in concrete is advantageous; such is not the case of a mixture of FeSi-75% and FeSi-50% or a mixture of FeSi-75% and CaSi.[2]

As is often the case in a plant in operation, a new modification that had not been foreseen during its construction is very difficult to implement and very costly. Hence, in some plants the total or large amounts of condensed silica fume produced has no potential value to the cement and concrete industry.

In view of this fact, before condensed silica fume can be used in concrete the following information should be available: (1) is the silica fume a blend of different types?; (2) type of alloy from which condensed silica fume is produced; (3) contaminants present in condensed silica fume.

The reader should note that most of the published data on the use of condensed silica fume in cement and concrete are related to condensed silica fume collected during the production of a silicon alloy containing at least 75% of silicon.

REFERENCES

1. Ferroalloy Statistics, Silicon Metal 8/84 P1 and Ferrosilicon 6/83 P2, Heinz H. Pariser, Alloy Metals & Steel, Market Research, Wesel, West Germany, 1983 and 1984.
2. Killin, A. M., Progress report air pollution control study of the ferroalloy industry, *Electric Furnace Proc.,* 31, 66, 1973.
3. Aïtcin, P.-C., Pinsonneault, P., and Roy, D. M., Physical and chemical characterization of condensed silica fumes, *Am. Ceram. J.,* 63, 1487, 1984.

Chapter 2

PHYSICAL CHARACTERISTICS AND PRODUCT VARIATION

In this chapter physical properties of different types of condensed silica fume are discussed with special attention to their variability with time from a given source.

I. PHYSICAL CHARACTERISTICS

Values of physical properties of condensed silica fume have been published by many authors but often without any details on the type of silica fume studied. In this chapter only clearly defined values will be discussed.

A. Color

Condensed silica fume varies in color from pale to dark gray (samples 1 and 2 in Figure 1); the exception is SiMn-CSF which is brown (sample 4), and condensed silica fume collected in a furnace equipped with a heat recovery system which is whitish (sample 3).

The carbon content and, to a lesser extent, the iron content seem to have a preponderant influence on the color of condensed silica fume. The darkest condensed silica fumes are those produced when wood chips are used in the raw feed of the furnace (sample 1 in Figure 1). The use of wood chips is mandatory to produce silicon metal, but optional when FeSi-90%-75% is produced. Hence, the fume from Si metal product (sample 1) is always darker than that from the production of FeSi-75% (sample 2).

A very simple method to check the carbon content of a particular condensed silica fume is to fill half a small bottle with condensed silica fume, for example, a bottle used to perform colorimetric tests on concrete sand — ASTM Standard C 40. Fill it up with water, agitate, and check the color of the suspension. The final color of the suspension taken instantaneously determines the amount of carbon.

When the condensed silica fume contains some carbon, this slurry takes on a pronounced black color (Figure 2, sample 1), and when the amount of carbon is under 2%, its color is generally dark gray (Figure 2, sample 2).

This very simple test is instructive because it also gives a clue as to how the use of a specific condensed silica fume will darken concrete color. If there are coarse particles in the fume, they will settle first and affect the homogeneity (Figure 3). It also indicates the presence or absence of chemicals that can go into the solution and color the water left over after the settling of the particles (Figure 4). These observations can be done 1 or 2 days after all particles have settled down.

B. Loose Bulk Specific Weight

Usually most condensed silica fumes are very light and very difficult to handle with the exception of SiMn. Table 1 presents loose bulk density of some specific condensed silica fumes.

ELKEM[1] and Popovic et al.[7] give values of 200 kg/m^3 for the loose bulk specific weight and 500 kg/m^3 when compacted.

SiMn fume, which is chemically very different from the other types of condensed silica fume, is denser. Its loose bulk specific weight is about the same as that of a very fine sand. When the color test is performed on SiMn in a small bottle, particles settle very rapidly, indicating that it is composed of coarser particles.

FIGURE 1.   Color of different types of condensed silica fume. (1) From the production of Si metal with wood chips; (2) from the production of FeSi-75% without wood chips; (3) from the production of FeSi-75% from a heat recovery furnace; (4) from the production of SiMn.

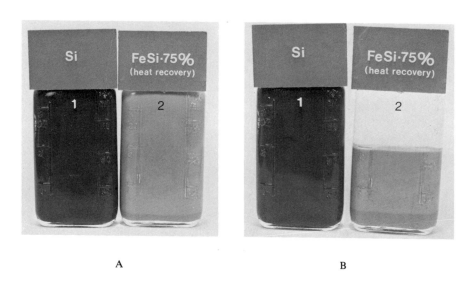

FIGURE 2.   Qualitative carbon content test: (A) immediately after shaking, (B) 24 hr later. (1) Condensed silica fume from the production of Si metal with wood chips; (2) condensed silica fume from the production of FeSi-75% without wood chips.

It is very difficult to compact condensed silica fume by vibration or by any other means because they seem to form large loose clusters that cannot be broken down easily (Figure 5).

The lightness has a strong economic impact on transportation and storage costs of condensed silica fume. For example, a cement tank designed to transport 35 t of Portland cement can handle no more than 10 to 12 t of loose condensed silica fume. The lightness of condensed silica fume has encouraged attempts to increase its specific weight in order to solve transportation problems.

C. Specific Gravity

Silica fume is composed mostly of vitreous silica particles, and its specific gravity is expected to be about 2.20, the most commonly accepted value for the specific gravity of vitreous silica. However, different impurities present in the glassy phase can alter

FIGURE 3. Close view of the settled particles.

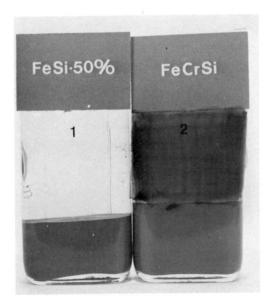

FIGURE 4. Color of the water above the settled particles. (1)
Condensed silica fume from the production of FeSi-50%; (2) condensed silica fume from the production of FeCrSi.

Table 1
AVERAGE LOOSE BULK SPECIFIC WEIGHT OF DIFFERENT
CONDENSED SILICA FUMES

	Si	FeSi-75%	FeSi-50%	FeCrSi	CaSi	SiMn	Ordinary Portland cement
kg/m³	260	230—405	310	330	530	535	800
lb/ft³	16	13—25	20	21	33	33	50

FIGURE 5.   Cluster of condensed silica fume particles.

Table 2
SPECIFIC GRAVITY OF DIFFERENT CONDENSED
SILICA FUMES[6]

Condensed silica fume type	Si	FeSi-75%	FeSi-50%	FeCrSi	CaSi	SiMn
Specific gravity	2.23	2.21—2.23	2.30	2.42	2.55	3.13

this value, for example, the presence of iron, magnesium, and calcium increases this value. The higher the amount of impurities, the higher the specific gravity. Table 2 presents some average values of the specific gravity of different types of condensed silica fume, of which SiMn silica fume is very different from the others.

Pistilli et al.[3,4] have found average values of 2.26 and 2.27 in their study of the uniformity of two Si and FeSi-75% condensed silica fumes.

D. Particle Shape

Condensed silica fume particles observed through an electron microscope appear in the form of a cluster of particles (Figure 5). When observed through a transmission electron microscope silica fume particles appear round (Figure 6). Under scanning electron microscopy (SEM) observation the particles appear to be spherical and smooth (Figure 6); the raspberry type surface described in a paper was in fact due to a poor preparation of the sample.

The particles have a wide range of sizes, but they are perfectly spherical. This spherical shape is the consequence of their origin, i.e., the condensation of a vapor. This process is currently used in the industry to make microspheres of different materials.

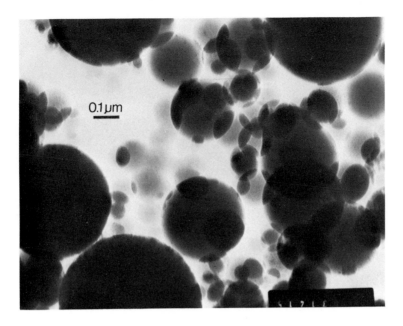

FIGURE 6.   Transmission electron micrograph of condensed silica fume particles.

Table 3

AVERAGE PARTICLE DIAMETER OF DIFFERENT CONDENSED
SILICA FUMES

Type of condensed silica fume	Si	FeSi-75%	FeSi-75% (heat recovery)	FeSi-50%	FeCrSi	CaSi	SiMn
Mean diameter (μm)[7]	0.18	0.26	0.23	0.21	0.18	—	—

E. Grain Size Distribution and Fineness

Very few publications have given complete grain size distribution of condensed silica fume particles.[1,8-10] It is not easy to determine the particle size distribution of such a fine material (Table 3). This also explains the wide variation reported in the literature (Figure 7).

The fineness of condensed silica fume can be evaluated by three different methods:

1.   The residue by sieving on a 45-μm sieve.[2-4]
2.   The direct measurement of the diameter of about 200 condensed silica fume particles.[6]
3.   The measurement of the specific area, which is an indirect measurement of the fineness. This is the most common method used.

1. 45-μm Residue

The residue on 45-μm sieve (no. 325) does not provide information on the grain size distribution of a particular condensed silica fume but rather on the amount of coarse impurities it contains, such as wood chips and quartz particles, or carbon or graphite particles from the burden or from the electrodes.

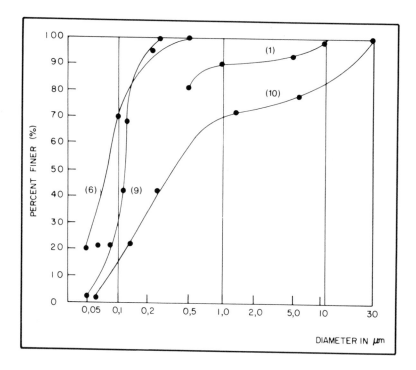

FIGURE 7.   Particle size distribution of condensed silica fume. The numbers in parentheses correspond to the references.

Generally the residue on a 45-μm sieve averages less than 5%. However, this value can be exceeded in some cases. ELKEM data[1] give a range of 0.3 to 3.5%, Pistilli et al.[3,4] found average values of 3.7 and 5.6% and Nebasar and Carette[2] 1.8 and 5.4% for a FeSi-75% and a Si fume, respectively.

Most of these large particles can be eliminated easily by a simple sieving on a 160-μm (no. 100) mesh. Condensed silica fume producers should make efforts to eliminate these undesirable coarse particles.

2. Grain Size Distribution by Electron Microscopy

Grain size distribution of condensed silica fume particles can be determined using an electron microscope. Using a special technique developed at The Pennsylvania State University, it has been possible to establish the grain size distribution of different types of condensed silica fume. In Figure 8 these results are presented in terms of numbers. It can be seen that the grain size distribution of condensed silica fume varies from one sample to the other, the finest particles originating from the Si condensed silica fume and the coarser, from the SiMn condensed silica fume.

Condensed silica fume from Si manufacture is expected to be the finest because the production of silicon metal is associated with the highest energy input and temperature. As the amount of iron increases in the alloy, the temperature of the furnace decreases and this is also the temperature at which fumes come out.

From the grain size distribution curve the average particle diameter has been calculated and is shown in Table 3. ELKEM[1] has published values presented in Table 4. The method of determination of these values was not indicated.

Condensed silica fume is a much finer material than Portland cement or other supplementary materials that are used in cement and concrete technology. Figure 9 shows

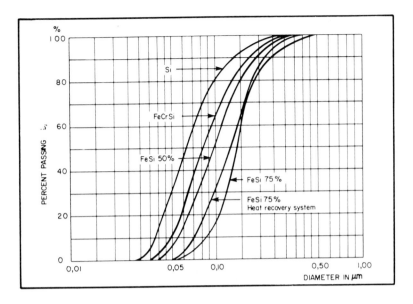

FIGURE 8.   Particle size distribution of different types of condensed silica fume particles.

Table 4
PARTICLE SIZE
DISTRIBUTION AS GIVEN BY
ELKEM[1]

Size distribution, free from coarse particles

+10 μm	1.5%
+ 5 μm	7%
+ 1 μm	10%
+ 0.5 μm	19%

the relative size of an average cement particle and average condensed silica fume particles. The extreme fineness of silica fume is responsible for its high reactivity. The observed difference in the reactivity of different types of silica fume can also be related to some extent to the differences in the size of condensed silica fume particles.

Condensed silica fume particles are approximately 100 times finer than other cementitious materials presently used by the cement and concrete industry.

F. Specific Surface Area
The specific surface area of a given powder is directly related to the size of its particles. For cement, fly ashes, and slags it is conveniently evaluated by the Blaine method and is expressed in m²/kg.

In some publications,[3,4,7] Blaine values are adopted for the specific surface area in spite of the fact that is is impossible to reach the 0.50 porosity level that is recommended by the standard for the measurement to be applicable.

Specific surface area of condensed silica fume is most commonly measured by nitrogen adsorption method (also known as the B. E. T. method, from Brunauer, Emmet, and Teller).

The specific surface area of a cement measured by nitrogen adsorption yields a dif-

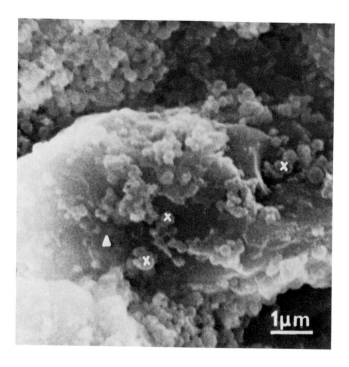

FIGURE 9.   Comparison of the size of a cement particle and a condensed silica fume particle: (Δ) cement particle, (x) condensed silica fume particles.

ferent value from that obtained with the Blaine apparatus. Generally, a cement that has a 300 to 400 m²/kg Blaine specific area is found to yield 1000 to 1500 m²/kg by nitrogen adsorption.

Most reported specific surface area values for condensed silica fume are between 13 and 30,000 m²/kg (13 to 30 m²/g). Nebesar and Carette[2] gave an average value of 20,000 and 17,200 m²/kg for 24 samples of an Si and FeSi-75% condensed silica fume. Values reported by ELKEM[1] lie in the range of 18,000 to 22,000 kg/m².

In the nitrogen adsorption method the amount, type, and fineness of the carbon present contributes to the surface area. As the carbon content can be related to the loss on ignition, the higher the loss on ignition, the higher the carbon content of condensed silica fume, and the higher the specific surface area. In order to interpret correctly the specific surface area, the LOI and carbon content values should also be provided.

The ratio of 100 (the average diameter of condensed silica fume to cement particles; see Figure 9) cannot be applied to specific surface area values because of the differences in the shape of cement and condensed silica fume particles. Two factors in condensed silica fume particles, i.e., spherical shape and perfect smoothness tend to minimize the specific surface area, whereas in cement particles the irregular shape and rough surface tend to increase the specific surface area.

The high specific surface area of condensed silica fume in concrete is responsible for high pozzolanicity and high water demand.

There exists a good relationship between the surface area values reported above and the results using the electron microscope. Because condensed silica fume particles are spherical, a simple relationship exists between the average diameter of the particles and the specific surface area. The average diameter of the microspheres and the specific surface area are related by the following simple formula[8]

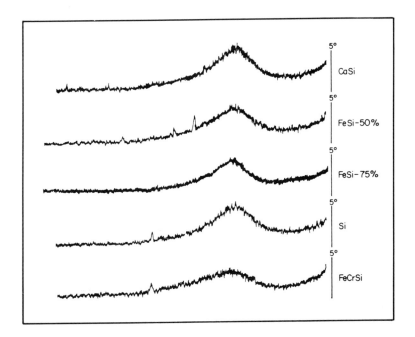

FIGURE 10.   X-ray diffractograms of different types of condensed silica fume.

$$D = \frac{6}{A \times ds}$$

where D is the average diameter in micrometers, A is the specific surface area expressed in m^2/g, and ds is the specific gravity of the particular condensed silica fume.

From this equation, it if found that a particular condensed silica fume having a specific gravity of 2.25 and a specific surface area of 20,000 m^2/kg ($20^2/g$) should be composed of microspheres having an average diameter of 0.13 μm, a reasonable value.

G. Amorphousness

X-ray diffractograms of samples of different types of condensed silica fume show them to be vitreous.[2,6] Popovic et al.[7] have found some SiC peaks, because SiC is an intermediate compound formed during the processing of silicon and the ferrosilicon. All the diffractograms exhibit a very wide scattering peak centered at about 4.4 Å, the most important peak of cristobalite (Figure 10). The FeCrSi halo is less well developed, probably because of the presence of a high amount of MgO.

When heated to 1100°C, most condensed silica fume crystallizes in the form of α-cristobalite, except FeSi-50% which crystallizes as enstatite, most probably due to the presence of a high amount of iron and magnesium oxide (Figure 11).

II. PRODUCT VARIATION

Condensed silica fume, being only a by-product, varies in its degree of purity and consistency because these are not major concerns for silicon or ferrosilicon production. On the other hand, they are not all unprocessed by-products like fly ashes. These are the by-products of a carefully controlled industrial process. The fabrication of silicon or ferrosilicon alloy in an electric arc furnace is carefully controlled just as the production of slag or pig iron. Hence, these types of industrial by-products are much more uniform than by-products rejected from unprocessed raw materials.

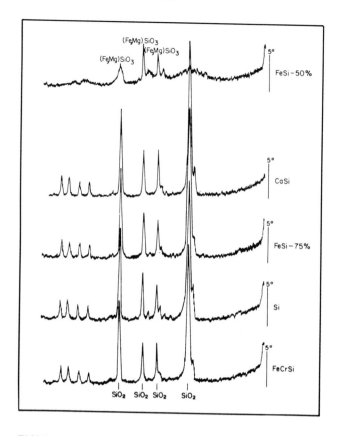

FIGURE 11.   X-ray diffractograms of different types of condensed silica fume after heating at 1100°C.

When processing a specific iron ore in a blast furnace, the slag composition is maintained within narrow limits for the production of the same type of pig iron (a stable rate which saves energy). This is done by adding fluxes in the raw feed so that the chemical composition of the slag varies only within a narrow range.

The same principle applies during silicon or ferrosilicon production. The silicon and ferrosilicon produced in an electric arc furnace have to comply with the ASTM or the customer's particular specifications. The only way to accomplish this is to very carefully control the composition of the raw feed to the electric arc furnace (quartz, carbon, and iron, for ferrosilicon). Thus, for a given type of silicon or alloy produced in a given furnace, the chemical composition of condensed silica fume is constant. However, a given furnace does not necessarily produce exactly the same type of silicon or silicon alloy year-round, so there can be slight variations in the chemical composition.

Only limited data have been published on the uniformity of condensed silica fume, though ELKEM has collected a great deal of information on this subject (Table 5).

Three publications give data on the chemical composition. The first was presented at the First International Conference on the Use of Fly Ash, Silica Fume, Slag and Other Mineral By-Products in Concrete held in Montebello, Canada in July 1983.[2] The papers by Pistilli et al.[3,4] are related to the variability in the composition of the two sources of condensed silica fume.

Nebesar and Carette[2] studied 96 samples, 48 from a silicon furnace and 48 from a FeSi-75% ferrosilicon furnace taken during a 6-month period. The frequency of sampling was maintained at a minimum of one sample for a maximum of 400 t of produc-

Table 5

CHEMICAL COMPOSITION OF THREE TYPES OF CONDENSED SILICA FUME

	Silicon 100	Silicon 90	Silicon 75
% SiO_2	94—98	90—96	86—90
% SiC	0.1—1.0	0.2—0.5	0.1—0.4
% C (total)	0.2—1.3	0.5—1.4	0.8—2.3
% Fe_2O_3	0.02—0.15	0.2—0.8	0.3—5.0
% Al_2O_3	0.1—0.4	0.5—3.0	0.2—1.7
% CaO	0.08—0.3	0.1—0.5	0.2—0.5
% MgO	0.2—0.9	0.5—1.5	1.0—3.5
% Na_2O	0.1—0.4	0.2—0.7	0.3—1.8
% K_2O	0.2—0.7	0.4—1.0	0.5—3.5
% P	0.03—0.06	0.03—0.04	0.03—0.04
% S	0.1—0.3	0.1—0.4	0.2—0.4
% TiO_2	0.001—0.02	0.02—0.06	0.02—0.06
% Mn	0.004—0.05	0.1—0.2	0.1—0.2
% Ni	0.001—0.002	0.005—0.04	0.02—0.04
% Co	0.001—0.002	0.05—0.06	0.05—0.06
% Cd	<0.0001	<0.0001	<0.0001
% Pb	0.001—0.002	0.005—0.007	0.005—0.006
% Cu	0.002—0.01	0.01—0.03	0.01—0.03
% Zn	0.006—0.015	0.006—0.015	0.006—0.015
% Mo	<0.005	<0.005	<0.005
% B	0.001—0.003	0.005—0.007	0.005—0.007
% LOI	0.8—1.5	0.7—2.5	2.0—4.0

tion for each furnace. The chemical composition was determined using an inductively coupled plasma optical emission spectrometer (ICP-OES) and specific surface area was determined by nitrogen absorption (single point B. E. T. method). Fineness (45 μm passing), pozzolanic activity index with Portland cement or with lime, and the water requirement were determined according to ASTM Standard Method C311-71.[5]

In their first paper Pistilli et al.[3] analyzed 32 samples originating from the same plant, but their sampling was done directly at the bag-house outlet so that it was composed of Si, FeSi-75%, or a mixture of the two fumes. Each sample represented 45 to 90 t of condensed silica fume. They were collected on consecutive days. In their second paper, Pistilli et al.[4] tested 30 samples from a U.S. plant producing only FeSi-75% alloy. These samples were collected daily and represented 45 to 90 t of condensed silica fume production.

The results of the above three studies are presented in Tables 6 and 8 for the average values, and in Table 7 and 9 for the standard deviations observed in each of the characteristics presented in Tables 6 and 8.

Based upon the results of the above studies, it can be concluded that the chemical composition and the physical characteristics of condensed silica fume from either silicon or ferrosilicon furnace of any one plant under study were reasonably uniform as analyzed by standard statistical techniques.

There are some differences, however, between the two types of condensed silica fume studied (Si and FeSi-75%). For example, the fume from the ferrosilicon furnace has higher amounts of iron, magnesium, and potassium oxides, but a lower amount of carbon in comparison with that produced in the silicon metal furnace. This is because iron is contributing some impurities to the raw mix and that the quartz used for FeSi-75% is not as pure as the one use to produce silicon. Also, the fume from the ferrosilicon furnaces showed a lower specific surface area but a higher fineness value with the

Table 6
AVERAGE CHARACTERISTICS OF FOUR CONDENSED SILICA FUMES

| Nature of CSF | Nebesar & Carette results (24 samples) | | Pistilli et al. results | |
	Si-CSF	FeSi-75%-CSF	32 samples Mixture of Si & FeSi-75%-CSF	30 samples FeSi-75%-CSF
Fineness by 45-μm sieve (% passing)	94.6	98.2	94	96.3
Specific surface area (m²/kg)	20,000[a]	17,200[a]	3,750[b]	5,520[b]
Pozzolanic activity index with Portland cement (%)	102.8	96.5	91.9	95.3
Water requirement (%)	138.8	139.2	140.1	144.5
Pozzolanic activity index with lime (MPa)	8.9	—	7.0	9.1
Specific gravity	—	—	2.27	2.26

[a] Nitrogen adsorption.
[b] Blaine permeability.

Table 7
STANDARD DEVIATIONS OF THE DIFFERENT CHARACTERISTICS OF THE FOUR STUDIED CONDENSED SILICA FUMES

| Nature of CSF | Nebesar & Carette results (24 samples) | | Pistilli et al. results | |
	Si-CAF	FeSi-75%-CSF	32 samples Mixture of Si & FeSi-75%-CSF	30 samples FeSi-75%-CSF
Fineness by 45-μm sieve (% passing)	4	1.5	1.7	4.5
Specific surface area (m²/kg)	2100	2700	680	785
Pozzolanic activity index with Portland cement (%)	5.1	13.7	10.0	4.0
Water requirement (%)	4.2	7.2	2.6	2.0
Pozzolanic activity index with lime (MPa)	0.8	—	0.8	0.9
Specific gravity	—	—	0.02	0.08

45-μm sieve. The higher specific surface area of the sample from the Si metal furnace can be related to the higher carbon content of the fume and finer size of the silica spheres.[6] The lower fineness determined by the 45-μm sieve for the silicon fume is due to the presence of some unburned wood particles. The use of a large amount of wood chips is mandatory for silicon production, while it is not so for ferrosilicon production. When the hot gases are sucked up from the furnace, they carry some incandescent wood chips that are supposed to be separated from the gases in "sparkle arrestors". Some small pieces of incandescent wood pass from these sparkle arrestors and are collected in the bag-house. From time to time, the bag-house sleeve filters have to be

Table 8

AVERAGE CHEMICAL COMPOSITION OF THE FOUR CONDENSED SILICA FUMES

Type of CSF	Nebesar & Carette results (24 samples)		Pistilli et al. results		
	Si-CSF	FeSi-75%-CSF	32 samples Mixture of Si-CSF & FeSi-75%-CSF	6 samples FeSi-75%	
SiO_2	93.7[a]	93.2	92.1	91.4	0.92
Al_2O_3	0.28	0.3	0.25	0.57	0.03
Fe_2O_3	0.58	1.1	0.79	3.86	0.41
CaO	0.27	0.44	0.38	0.73	0.08
MgO	0.25	1.08	0.35	0.44	0.05
Na_2O	0.02	0.10	0.17	0.20	0.02
K_2O	0.49	1.37	0.96	1.06	0.05
S	0.20	0.22	—		
SO_3	—	—	0.36	0.36[b]	0.16[b]
C	3.62	1.92	—		
LOI	4.4	3.1	3.20	2.62[b]	0.42[b]

[a] Calculated.
[b] From 30 samples.

Table 9

STANDARD DEVIATION OF THE OXIDE CONTENT OF THE FOUR CONDENSED SILICA FUMES

Nature of CSF	Nebesar & Carette results		Pistilli et al. results	
	Si-CSF	FeSi-75%-CSF	Mixture of Si and FeSi-75% CSF	FeSi-75%-CSF
SiO_2	3.8	1.7	1.3	0.9
Al_2O_3	0.13	0.20	0.12	0.03
Fe_2O_3	2.26	0.90	0.70	0.41
CaO	0.07	0.34	0.11	0.08
MgO	0.26	0.29	0.10	0.05
Na_2O	0.02	0.06	0.04	0.02
K_2O	0.24	0.45	0.22	0.05
S	0.16	0.06	—	—
SO_3	—	—	0.10	0.16
C	0.46	1.15	—	—
LOI	1.5	0.9	0.45	0.42

replaced because they contain too many holes due to the burning of the filtering material.

Any two well-identified FeSi-75% condensed silica fumes will have almost the same chemical analysis and physical properties, showing that a particular type of alloy generates a well-defined condensed silica fume.

In spite of the slight differences between the Si and FeSi-75% fume, the pozzolanic index activity index for the two types of fumes is not significantly affected.

The above results indicate that a given condensed silica fume can be considered as quite a uniform material resembling blast furnace slags rather than fly ashes.

As mentioned already, the consistency of values for a particular alloy is recognized. If the type of alloy changes, so do the characteristics of the corresponding condensed silica fume. It is desirable, therefore, that condensed silica fume users know any changes in the source of the raw feed used for the furnace or changes in the nature of the alloy produced by a plant.

Some condensed silica fumes may not be as uniform as the types mentioned above, particularly when the silicon or ferrosilicon plant is equipped with only one bag-house in which the fumes from different furnaces producing different types of alloys are collected.

REFERENCES

1. ELKEM SILICA Technical Bulletin, ELKEM — SPIGERVERKET a/s, Fiskaa Verk, P.O. Box 40, 4620 Vaayobygd, Norway, 1980.
2. Nebesar, B. and Carette, G. G. Variations in the Chemical Composition, Surface Area, Fineness and Pozzolanic Activity of a Canadian Condensed Silica Fume, Project MRP-3.6.0.4.65, CANMET, Ottawa, Canada.
3. Pistilli, N. F., Rau, G., and Cechner, R., The variability of condensed silica fume from a Canadian source and its influence on the properties of Portland cement concrete, *Cement, Concrete Aggregates,* 6, 33, 1984.
4. Pistilli, N. F., Wintersteen, R., and Cecnner, R., The uniformity and influence of silica fume from a U.S. source on the properties of Portland cement concrete, *Cement, Concrete Aggregates,* 6, 120, 1984.
5. Sampling and Testing Fly Ash or Natural Pozzolans for Use as a Mineral Admixture in Portland Cement Concrete, ASTM C311-71, *Annual Book of ASTM Standards,* Part 14, *Concrete and Mineral Aggregates,* American Society for Testing and Materials, Philadelphia, 1985.
6. Aïtcin, P.-C., Pinsonneault, P., and Roy, D. M., Physical and chemical characterization of condensed silica fume, *Am. Ceram. Bull.,* 63, 1487, 1984.
7. Popovic, K., Ukraincik, V., and Djurekovic, A., Improvement of mortar and concrete durability by the use of condensed silica fume, *Durability Build. Mater.,* 2, 171, 1984.
8. Aïtcin, P.-C., *Condensed Silica Fume,* Les Editions de l'Université de Sherbrooke, Sherbrooke, Quebec, Canada, 1983, 16.
9. Jenkins, R. D., Potential Utilization and Disposal of Particulate Materials Captured from a Silicon Metal Furnace, Globe Metallurgical Division Interlake Inc., Beverly, Oh., 1973, 1.
10. Kolderup, H., Particle size distribution of fumes formed by ferrosilicon production, *Air Pollut. Control Assoc. J.,* 27, 127, 1977.

Chapter 3

HANDLING AND TRANSPORTATION

I. INTRODUCTION

Unprocessed condensed silica fume, being very fine with a low bulk specific weight, 200 to 300 kg/m³, is not easy to handle. This is probably one of the several reasons why it was not used earlier, although its high efficiency as a pozzolan has been known for more than 30 years. There was no interest in collecting, transporting, and marketing such a product having a low economical value.[1,2]

II. STORAGE

For many years silicon and ferrosilicon producers did not make an effort to collect condensed silica fume because it was easier and more economical to dispose of it into the atmosphere. This attitude toward a by-product was not unique to the silicon and ferrosilicon industries; it existed in other industries. However, due to an uncontrolled increase of pollution in the ecosystem, very tight environmental regulations were recently enforced in most industrialized countries, and the producers of condensed silica fume were asked to collect the silica fumes and not allow them to escape into the atmosphere.

Presently, most of the silicon and ferrosilicon plants in developed countries collect the fume in a sophisticated dedusting system, working more or less like a big vacuum cleaner. This unit is generally known as the bag-house in silicon and ferrosilicon plants (Figure 1).

The introduction of a dedusting system to existing plants was not easy, not because it was technically difficult but rather due to the economics of building and operating it. This additional expense would have reflected in the production cost of the only saleable product, the silicon or ferrosilicon in a depressed market.

When the decision to add a bag-house to a plant was taken by the silicon and ferrosilicon industries, the reaction was that condensed silica fume was only a waste of no use. Generally, it is not a difficult technical problem to handle or store a liquid or a solid, even if it is hazardous, but with condensed silica fume the problem is that it defies gravitational law, and hence it is very difficult to contain it. Even if the containment of condensed silica fume can be solved by using very tight connections between the different parts of the dedusting and handling system, because of its low specific weight (only 10% of its apparent volume is composed of solid particles), very large storage facilities have to be designed to store it temporarily in a plant (Figure 2).

III. TRANSPORTATION

After containment and collection the next problem is transportation. Different approaches have been taken depending on local considerations. In some plants condensed silica fume is pelletized (Figure 3) in pelletizing discs or drums such as those used to pelletize iron ore. The drum pelletizer seems to work better with such a dusty fine powder. Water is used as a binder. The hardened pellets are disposed of in dumps like any ordinary granular material.

In some other plants condensed silica fume is introduced in a ready-mix truck, half-filled with water, and agitated to a more or less viscous slurry. This slurry is trans-

FIGURE 1.   Bag-house in a silicon plant in Bécancour, Quebec, Canada.

FIGURE 2.   Temporary storage facility for condensed silica fume
in Bécancour, Quebec, Canada.

FIGURE 3.   Pellets of condensed silica fume.

FIGURE 4.   Airtight truck box used to transport condensed silica
fume to a dump.

ported to a nearby pond for settlement. In still some other plants special large airtight
truck boxes are filled with condensed silica fume and dumped into a disposal pond
(Figure 4).

It was in Norway, where there are a number of silicon and ferrosilicon producers,
that most early efforts were made to transport and handle condensed silica fume for
use in the cement and concrete industries.

Different methods of packaging are adopted by companies. For example, ELKEM[3]
proposed the conditioning of condensed silica fume in the following forms:

FIGURE 5.   Transporting condensed silica fume in a large tank truck.

Bulk   in tank truck   (Figure 5)
Big bags   400—500 kg (compacted)   (Figures 6 and 7)
Big bags   300—350 kg (not compacted)
PVC bags   25 kg (not compacted)
              (also shrink wrapped on pallets)
              (Figure 8)

Pellets with 25 to 30% water may also be delivered in cases, provided the water content does not matter. Cement may be added to the pellets to achieve high strength.

Elkem-silica may also be delivered in a slurry form, containing 50% Elkem-silica and 50% water. This slurry is very stable and easy to pump and transport, e.g., in tankers. Cargoes of up to 1000 t of slurry may be loaded.

Recently an admixture company has proposed supplying condensed silica fume in a slurry form under the trade name of Force 10,000. It is a slurry containing 50 to 60% of solid particles.[4]

In the Quebec province of Canada, all condensed silica fume presently used by the concrete industry is transported in bulk. Concrete producers use regular cement tankers or have developed large airtight truck boxes to transport condensed silica. The loading and unloading of the cement tank takes longer when filled with condensed silica fume, but can be achieved using low pressure.[5] Occasionally unloading difficulties may occur. A cement tanker that normally contains 30 to 35 t of cement may contain only 10 to 12 t of condensed silica fume. As the handling distance in Quebec never exceeds 100 km, the transportation costs are not excessive. All that is involved in using condensed silica fume is to have a new silo and to learn how to unload a cement tanker filled with condensed silica fume.

IV. DENSIFICATION

In parallel to the marketing efforts, a constant concern has been to densify condensed silica fume in order to transport and handle it more easily and make it more attractive to the cement and concrete industries.

FIGURE 6.   Filling a large bag with condensed silica fume.

FIGURE 7.   Transport of condensed silica fume in large bags.

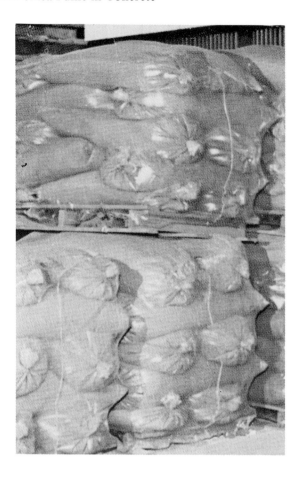

FIGURE 8.   Condensed silica fume in 25 kg PVC-bags shrink
wrapped in pallets.

The slurry process may be considered a process of densification, but in order to transport 1 t of condensed silica fume it is necessary to transport 1 t of water. One technical advantage of this process is that condensed silica fumes are very well dispersed when introduced in concrete.

It has also been found that condensed silica fume can be densified in the form of a coarse powder such as cement. This powder has better packing characteristics than the bulk condensed silica fume.

It was accidentally found that by maintaining condensed silica fumes in a slow motion in a collecting silo they were transformed (in the absence of any binding agent) into micropellets (Figure 9) with a specific weight twice that of virgin condensed silica fume.[6] Bulk specific weight of 500 kg/m³ can be reached with this process. The resulting micropellets can be easily transported using cement tanks. In this process specific weight is not different from that using slurry process; however, transportation of 1 t of water for every metric ton of fume is avoided. Moreover, the transportation and handling of a dry powder is more familiar to the cement and concrete producers than the transportation of a slurry. With the powder, under adverse winter conditions of the northern U.S. and Canada no special protection is necessary as would be needed for a slurry.

FIGURE 9. Micropellets of condensed silica fume.

FIGURE 10. Densified condensed silica fume according to the University of Sherbrooke patent.

Some plants in the world are equipped with composing micropelletizing systems as
described earlier, but the main problem with this is that it is a slow process. It requires
several days to achieve the desired level of densification. As this process cannot be
speeded up, the only solution is to build a series of big silos and let condensed silica
fume reach the desired level of densification; this requires a large capital investment.

One technical advantage of the use of micropellets is that when they are introduced
into concrete they are well dispersed unlike large micropellets obtained from a drum
pelletizer. In the latter case there is always a certain percentage of condensed silica
present in the hardened concrete in the form of unreacted small pellets. These small
pellets represent the hard cores of dense, hard, large pellets.

Condensed silica fume can also be transformed into a coarser form in the same range
of fineness as Portland cement (Figure 10), according to a patented process developed
at the University of Sherbrooke.[7] With this process a specific weight of 400 kg/m³ can

be reached very rapidly. When densified by this process, condensed silica fume is as easy to handle as cement and is very well dispersed in concrete with or without a superplasticizer. This process has been tried successfully in a pilot operation, but is not presently used in a full-scale operation.

REFERENCES

1. Jenkins, R. D., Potential Utilization and Disposal of Particulate Materials Captured from a Silicon Metal Furnace, Globe Metallurgical Division, Interlake Inc., Beverly, Oh., 1973, 1.
2. Hällgreen, E., Disposal of Fume from Submerged Arc Furnaces in Europe, *Electric Furnace Conf. Iron Steel Soc. AIMR,* 33, 76, 1975.
3. ELKEM-SILICA, Technical Bulletin, ELKEM—SPIGERVERKET a/s Fiskaa Verk, P.O. Box 40, 4620 Vaayobydg, Norway, 1980.
4. 10,000 Force, W. R. Grace Technical Bulletin, 1985, available from W. R. Grace & Co., Cambridge, Mass.
5. Skrastins, J. I. and Zoldners, N. G., Ready-Mixed Concrete Incorporating Condensed Silica Fume, Proc. 1st Int. Conf. on the Use of Fly Ash, Silica Fume, Slag and Other Mineral By-Products in Concrete, Montebello, Quebec, Canada, July 1983 (available as ACI SP 79, Vol. 2, pp. 813-830, Malhotra, V. M., Ed., American Concrete Institute, Detroit).
6. Bruff, W. and Novosad, S., Balling of silica dust by mechanical handling, in *Powder Technology,* Elsevier Sequoia SA23, Elsevier, Amsterdam, 1979, 273.
7. Aïtcin, P. C., Pinsonneault, P., and Fortin, R., Agglomerated Volatilized Silica Dust, U.S. Patent 4,384,896, May 24, 1983.

Chapter 4

REACTIONS IN THE CEMENT-SILICA FUME-WATER SYSTEM

I. INTRODUCTION

As with other industries, progress in concrete technology should necessarily take into account the need for conserving national resources and the environment and proper utilization of energy. Consequently, there is a constant demand for the utilization of wastes and by-products. Some of the ways in which these materials are used are as raw material in the manufacture of cement clinker, manufacture of cement, as aggregates with or without processing, and as admixtures. Many types of materials such as fly ash, slag, spent oil shale, mining and quarrying wastes, rice husk ash, and silica fume may be used for making concrete.

Burning of rice husks produces an ash containing 80 to 95% SiO_2, 1 to 2% K_2O, and carbon. In such ashes silica is present in an amorphous state. The surface area can be as high as 50 to 60×10^3 m^2/kg. The rice husk-Portland cement mortars show much higher strengths than Portland cement mortars. For example, the ash:cement (70:30) mixture exhibits a compressive strength of 31.9, 45.5, and 58.7 MPa at 3, 7, and 28 days, respectively, and the corresponding strengths for a straight cement mortar are only 22.4, 32.5, and 42.4 MPa.[1] Silica fume, being similar to rice husk in SiO_2 content and having a high surface area can also be used in cements to obtain high strengths. The strength development, as with other physical and mechanical properties, is related to the hydration kinetics in cement-silica fume mixtures. Silica fume, in addition to increasing the rate of hydration of C_3S, combines with the lime formed as a reaction product of cement hydration. This chapter discusses the chemical aspects of the reaction of cement containing silica fume.

II. HYDRATION OF TRICALCIUM SILICATE

A. Rate of Hydration

Only a limited amount of work has been carried out on the influence of silica fume on the hydration of C_3S. Finely divided silica-based materials can serve as models for predicting the behavior of silica fume. Stein and Stevels[2] studied the effect of SiO_2 in the form of quartz (low surface area) or aerosil (surface area of about 200×10^3 m^2/kg) by following changes in the electric conduction of solution, heat effects, and microstructure and concluded that the early hydration of C_3S is accelerated by aerosil but not by quartz. Thus, it becomes evident that high surface area silica acts as an accelerator for C_3S hydration. The importance of surface area can be illustrated by using $CaCO_3$ as an example. Although apparently inert in the C_3S-H_2O system, high surface area $CaCO_3$ can accelerate C_3S hydration. Figure 1 demonstrates the effect of finely divided $CaCO_3$ of surface area 6×10^3 m^2/kg on the hydration of C_3S.[3] Addition of 5 to 15% $CaCO_3$ increases the degree of hydration of C_3S. The accelerating effect is particularly significant in the first few days.

Kurdowski and Nocun-Wczelik[4] mixed amorphous silica with C_3S at C_3S/SiO_2 molar ratios of 0.4 to 3.33 and followed the rate of hydration up to 24 hr. Figure 2 shows the rate of heat evolution of C_3S containing different molar proportions of SiO_2. A hump with a peak at about 8 hr occurs with C_3S containing no SiO_2. At a C_3S/SiO_2 molar ratio of 3.33 the initial peak (below 60 min) and the second peak are intensified, suggesting acceleration effects. As the amount of added SiO_2 is increased the first peak is

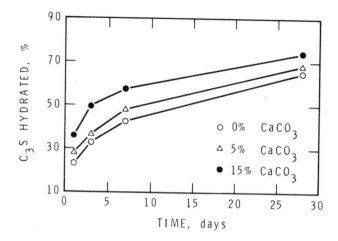

FIGURE 1.   Influence of CaCO₃ on the hydration of C₃S.

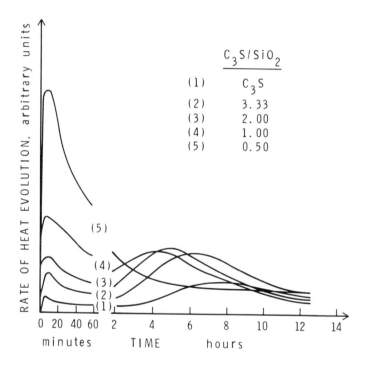

FIGURE 2.   Heat evolution curves for samples of C₃S containing silica.

intensified and the induction period is decreased. At low C_3S/SiO_2 ratios the second peak seems to disappear.

The rate of hydration may also be followed by the amount of C_3S remaining during hydration (Figure 3).[4] From the results obtained with C_3S/SiO_2 ratios of 1 and 0.4, it is evident that the addition of SiO_2 accelerates the disappearance of C_3S. Higher dosages of SiO_2 have a more effective accelerating effect. The reactivity of silica can also be assessed from the estimation of the amount of $Ca(OH)_2$ formed at different times. For example, the amount of $Ca(OH)_2$ formed decreases from 15% with 0% SiO_2 to 0% in the presence of SiO_2 (CaO/SiO₂ ratio of 0.4). This is due to the reaction of SiO_2 with the $Ca(OH)_2$ formed by the hydration of C_3S.

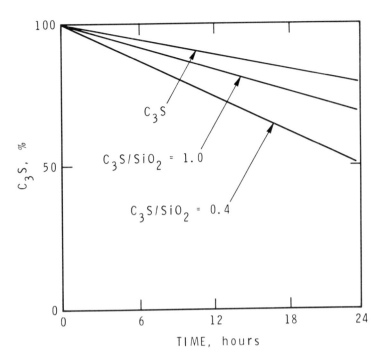

FIGURE 3.   The influence of silica on the C_3S hydration.

Similar conclusions may be drawn from the work of Ogawa et al.[5] who used white siliceous clay containing 87.8% SiO_2. This silicate had a surface area of 6.7×10^3 m²/kg. Estimation of the degree of hydration of C_3S showed that at day 1 C_3S hydrated to an extent of 40% without the additive and in the presence of the white clay the degree of hydration was 80%. Figure 4 estimates the amount of $Ca(OH)_2$ formed at different times of hydration. Although at day 1 the amount of $Ca(OH)_2$ is greater in the sample hydrated with the siliceous addition, at 7 days there is substantially more $Ca(OH)_2$ in the product obtained from pure C_3S. The lower amounts of $Ca(OH)_2$ at later periods of hydration may be attributed to the reaction of SiO_2 with the $Ca(OH)_2$ formed as a hydration product of C_3S.

A systematic investigation of the effect of a high surface area silica (200×10^3 m²/kg) on the hydration of alite has recently been reported by Wu and Young.[6] The rate of heat evolution within the first 2 hr in the C_3S-SiO_2 mixtures containing 0, 6, 14, 21, 28, and 34% SiO_2 is shown in Figure 5. The intensity of the peak increases as the amount of added SiO_2 is increased and a shoulder appears at higher SiO_2 contents. In contrast to the results of Kurdowski and Nocun-Wczelik[4] the second main peak (not shown in the figure) was never obliterated. The peak area in the sample containing 34% SiO_2 is about five times greater than that of C_3S containing 0% SiO_2. The increase in the intensity of the peak may be attributed to the acceleration of C_3S hydration. The shoulder represents a rapid reaction between SiO_2 and CH released from the alite. The early decrease in the (OH) concentration in the solution phase confirms this conclusion. It was also observed that the initial peak area increased as the surface area of added SiO_2 was increased.

Table 1 compares the degree of hydration of $C_3S(\alpha)$, C/S ratio of C-S-H, and CH formed at different times in the C_3S-SiO_2 mixtures containing 0, 14, and 34% SiO_2. The increase in the degree of hydration in terms of the consumption of C_3S is evident.

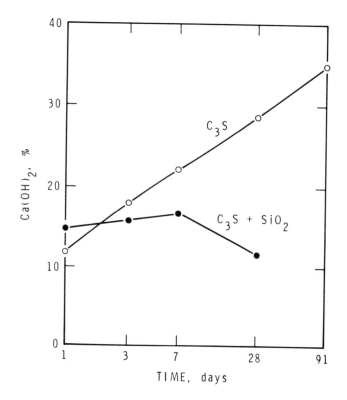

FIGURE 4.   Amount of Ca(OH)$_2$ in C$_3$S-siliceous material paste.

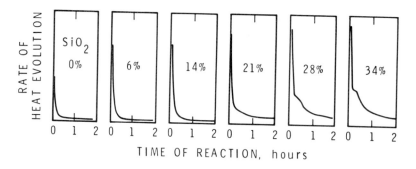

FIGURE 5.   Rate of heat evolution of C$_3$S-Cab-o-Sil mixtures containing various amounts of SiO$_2$.

The amount of CH decreases as the amount of SiO$_2$ is increased due to the reaction between SiO$_2$ and CH formed during hydration.

It must be emphasized that the variability in results reported by different workers is expected because of the difference in the types of silica, C$_3$S, water/solid ratios, and experimental procedures. Silica fume with a surface area of about 20×10^3 m^2/kg may be less reactive than that with an area of 200×10^3 m^2/kg but more reactive than that having a value of 6×10^3 m^2/kg.

B. CaO/SiO$_2$ Ratio of C-S-H

The CaO/SiO$_2$ ratio of the C-S-H product may be modified by the addition of SiO$_2$ to hydrating C$_3$S. There are certain limitations involved in the determination of CaO/

Table 1

ANALYSIS OF ALITE HYDRATION PRODUCTS IN THE PRESENCE OF HIGH SURFACE AREA SiO$_2$

Time	0% SiO$_2$			14% SiO$_2$			34% SiO$_2$		
	α	Ca(OH)$_2$	C/S	α	Ca(OH)$_2$	C/S	α	Ca(OH)$_2$	C/S
1 hr	—	—	—	—	—	—	16	1.0	1.46
6 hr	—	—	—	23.7	4.3	1.76	—	—	—
8 hr	—	—	—	—	—	—	57	1.8	1.33
14 hr	—	—	—	30.5	12.6	1.56	—	—	—
1 day	19.6	9.7	1.47	52.8	16.5	1.48	65.4	3.2	1.20
3 days	36.6	11.8	2.01	64.6	9.0	1.52	72.3	1.7	1.16
14 days	50.4	17.5	1.93	79.9	8.7	1.47	84.4	2.6	1.12
28 days	64.6	27.2	1.64	85.0	10.0	1.49	88.9	—	—

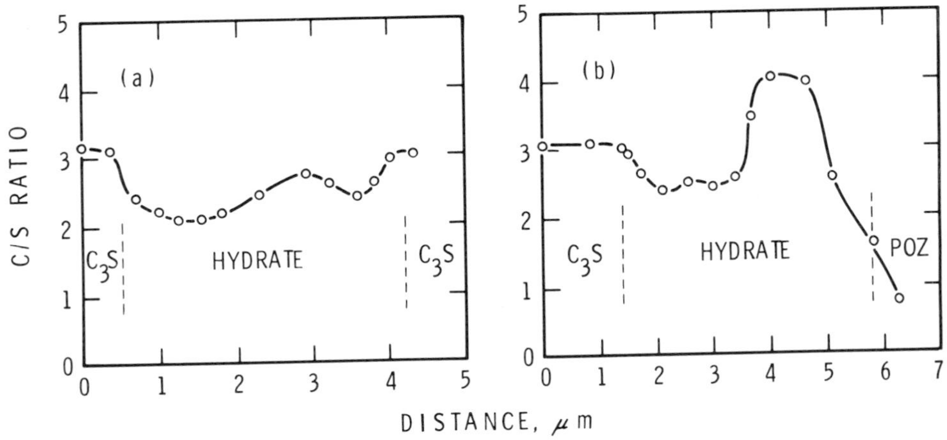

FIGURE 6. CaO/SiO$_2$ ratios at different spots in the C$_3$S-SiO$_2$ pastes. (a) Between C$_3$S particles and (b) between C$_3$S and white siliceous pozzolan.

SiO$_2$ ratio and, hence, the reported values vary. It is reported[4] that in the absence of the additive at a W/S ratio of 10, C$_3$S yields a C-S-H product with a CaO/SiO$_2$ ratio of 0.8 at 24 hr and this ratio is decreased to 0.33 and 0.36 at C$_3$S/SiO$_2$ ratios of 1 and 0.4, respectively. Using microanalytical methods, Ogawa et al.[5] determined the CaO/SiO$_2$ ratio from the edge of the C$_3$S particle through to the siliceous addition. Figures 6a and 6b illustrate that this ratio depends on the distance of any point from the edge of C$_3$S. At 3 days of hydration the ratio decreases from 3.0 to 2.5 to 3.0 in the zones of the hydrated product. In the mixture containing C$_3$S and the siliceous pozzolan this ratio is higher than that in hydrated C$_3$S (containing no siliceous additive). The C$_3$S grains are surrounded by 2-μm thick hydrate with a CaO/SiO$_2$ ratio of 2.5. Near the pozzolan there is a lime-rich hydrate. At the surface of the pozzolan the ratio is less than 1. These results would indicate that even in the C$_3$S-pozzolan mixture, in the area close to C$_3$S the course of hydration is substantially the same as in the C$_3$S used alone. At areas farther away from the C$_3$S particle, a low CaO/SiO$_2$ ratio product is evident from the pozzolanic reaction.

The C/S (cement/silica) ratio of the C-S-H product obtained by different workers is expected to show some variation because not only are the materials and procedures

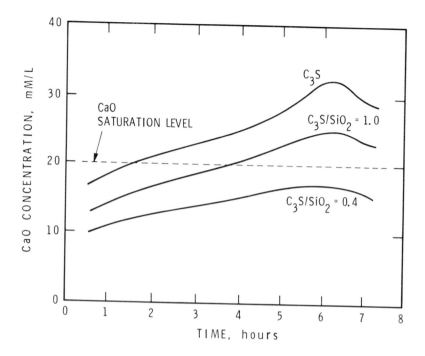

FIGURE 7.   Concentration of CaO in the liquid phase as a function of time.

different but uncertainties also exist in the calculations. The calculation of the C/S ratio involves determination of unreacted CaO and SiO_2 and of the CH formed during hydration. The amount of C-S-H formed is obtained by difference. There is, however, a general tendency for the C/S ratio of the C-S-H product to decrease as the amount of added SiO_2 is increased. Three kinds of C-S-H seem to be formed in the C_3S-SiO_2-H_2O system. One kind forms directly from the hydration of C_3S. The second type with a lower C/S ratio forms by the reaction of SiO_2 and CH (formed during C_3S hydration) and the third type forms from the reaction between SiO_2 and C-S-H.[6]

C. Mechanism

The complete mechanism of the accelerating effect of SiO_2 on C_3S hydration has not been worked out. One approach involves determining the concentration of Ca^{2+} ions in the solution phase of the hydrating C_3S (Figure 7).[4] The concentration of Ca^{2+} increases rapidly and reaches the supersaturation level and is above this level for at least 6 hr in the C_3S sample containing no addition. In the C_3S sample containing SiO_2 ($C_3S/SiO_2 = 0.4$) the Ca^{2+} concentration does not reach the supersaturation level. It is thus possible that transportation of Ca ions from the grain surface to the solution phase, influenced by silica, may determine the reaction rate. The rate of C_3S hydration is considered as depending on the consumption of Ca^{2+} from the solution phase forming $Ca(OH)_2$ and causing C-S-H precipitation. In the liquid phase around C_3S grain the existence of a quasi-stationary supersaturation of Ca^{2+} ions is envisaged. The diffusion of Ca ions from this supersaturated layer will be controlled by the concentration gradient that depends on the concentration of the supersaturated layer and the concentration in the bulk solution with respect to $Ca(OH)_2$. By the addition of SiO_2 a C-S-H product with low CaO/SiO_2 ratio is formed and the concentration of Ca ions in the bulk solution is reduced. This will affect the quasi-stationary supersaturated layer around the grains, resulting in an acceleration of C_3S. In this mechanism, nucleating effect is of no importance.

According to Stein and Stevels,[2] the accelerating effect of high surface area SiO_2 is explained by the formation of a more permeable C-S-H layer around the C_3S surface.

In the mechanism proposed by Ogawa et al.,[5] the silica surface acts as a preferable precipitation site and not the hydrate. In addition, SiO_2 adsorbs Ca ions, lowers Ca^{2+} concentration and promotes dissolution of C_3S.

The mechanism of hydration of C_3S in the presence of silica has also been elaborated by Wu and Young.[6] It envisages a sort of silica layer forming around C_3S particles. This is possible because of the enormous surface area of SiO_2 compared to that of C_3S. Upon contact with water, Ca and OH ions released by C_3S have to pass through this layer to enter the bulk solution, Ca and OH ions react with SiO_2 to form C-S-H, and delay the increase in Ca and OH ions in the bulk solution. Thus, the diffusion of Ca ions into the SiO_2 surface becomes the rate-controlling step. At higher W/S ratios the ions encounter a small number of SiO_2 particles in the vicinity of the C_3S surface. At a W/S = 1.4 , 30% CH is consumed whereas at W/S = 10, only 20% CH is consumed. Also, in the early stages C-S-H formation occurs on the high surface area, SiO_2 surface, rather than on the C_3S surface. This means the C_3S surface will not have a usual barrier, thus promoting the accelerated dissolution of the ions.

D. Polymerization

It is known that when C_3S is hydrated the hydrate is polymerized. In a fully hydrated cement paste, Si exists as monomer, dimer, linear trimer, and polymer in amounts of 9 to 11, 22 to 30, 1 to 2, and 44 to 51%, respectively. Polymer in this context includes all species except monomer and dimer.[1]

The degree of polymerization in C_3S containing SiO_2 has been followed using the trimethyl silyl (TMS) derivative of C-S-H according to methods developed by Tamas et al.[7] Monomeric silicate is represented by the unreacted C_3S. The polysilicate contents of C-S-H formed at different times of hydration are plotted in Figure 8.[6] The polysilicate content increases gradually to 40% at lower SiO_2 contents (composed of penta- and octameric species). At higher SiO_2 contents the polysilicate content reaches 80%. The highly polymerized C-S-H with low C/S ratio is formed by the reaction of silica with C-S-H, especially at higher SiO_2 contents. The implications of the polymerization in terms of mechanical and physical properties have yet to be resolved.

III. HYDRATION OF CEMENT

A. Initial Stages

Physical and chemical changes occur as soon as cement comes into contact with water. Within 15 to 30 min, the aluminate components react and Ca and OH ions are released into the solution. The rheological behavior of the fresh paste determines its initial physico-mechanical properties. The rheology of cement paste is followed by determining the deformation behavior under stress. These studies yield important insight into the role of admixtures and additions such as silica fume in cement. A viscometer is used to measure yield stress and viscosity. The behavior of cement paste can be described by the equation $\gamma = \gamma_y + \mu_p \gamma$ where γ = shear stress, γ_y = yield value, μ_p = plastic viscosity, and γ = shear rate.

Addition of silica fume to cement increases the yield value and plastic viscosity. In view of this, in cement-silica fume systems, a plasticizer has to be added for deflocculating the paste. The higher the yield value, the larger the surface area of the silica fume addition. In Figure 9, the mix containing a silica fume of surface area 27×10^3 m^2/kg shows a higher yield value than that containing a lower surface area silica fume $(17.7 \times 10^3 \ m^2/kg)$.[8,9] The water demand and the effect of silica fume on the workability of concrete are discussed in another chapter.

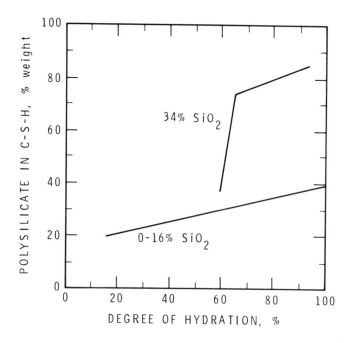

FIGURE 8.   Amount of polysilicate in C-S-H formed in C₃S paste with
and without silica (the remainder is dimeric silica).

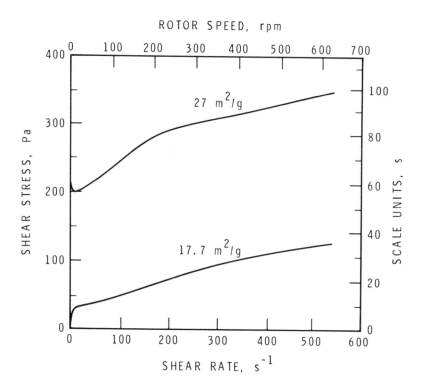

FIGURE 9.   Rheological behavior of cement pastes containing silica fume of different
surface areas.

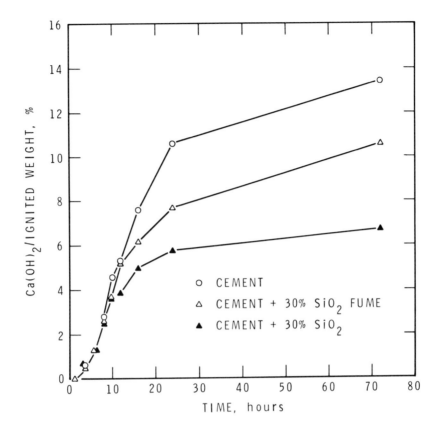

FIGURE 10. Change in Ca(OH)₂ content at early times for cement and cement blends.

B. Estimation of Hydration Products

The effect of silica fume on the kinetics of hydration of cement is not easy to follow because not only does silica influence the rate of hydration, it is also consumed by the reaction products. The variation in results reported in the literature may also be attributed to the differences in the reactivity of the materials, techniques, and preparation methods.

1. Calcium Hydroxide

The sequence of hydration of cement has been followed by determining the amount of the products formed at different times.

The amount of $Ca(OH)_2$ formed at different times by the addition of 30% silica fume (21×10^3 m²/kg) to cement was compared with that formed using a relatively coarsely ground silica (passing 100 mesh; 75 μm). The results are plotted in Figure 10.[10] The amounts of $Ca(OH)_2$ are based on ignited weight. The three curves are similar up to 8 hr, but at later periods cement paste containing no additives shows higher CH contents. This would indicate that both silica fume and ground silica interact with CH formed during hydration. These results also suggest that silica fume of greater surface area reacts more efficiently with CH. This clearly shows that silica fume is a good pozzolan.

Ono et al.[11] investigated the effect of silica fume (22.8×10^3 m²/kg) on the production of lime (as CaO) in the paste hydrated at a W/S ratio of 0.23 (Figure 11). In the figure, CaO is reported on an ignited basis. The amount of CaO formed in these pastes

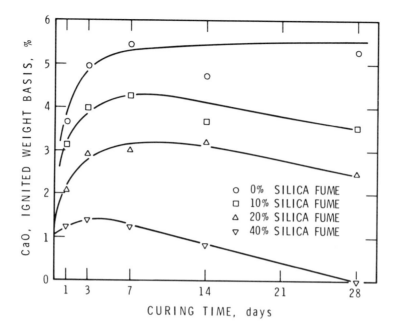

FIGURE 11.   Amount of Ca(OH)₂ (as CaO) in cement containing different amounts of silica fume.

is lower than that formed at the higher W/S ratio of 0.6 reported in Figure 10. Afte an initial increase in the amount of CaO, the amount decreases in all samples containing silica fume. At higher dosages of silica fume there is a larger decrease in CaO, and in the samples containing 40% SiO_2 no lime could be detected at 28 days. These results confirm that silica fume reacts with lime formed during hydration and the amount of free lime depends on the rate of hydration of Portland cement and the amount and surface area of silica fume.

2. Nonevaporable Water

The determination of the nonevaporable water content offers another method of following hydration. The nonevaporable water includes all water combined chemically with $Ca(OH)_2$, and the silicate and aluminate phases. Figure 12a compares the non-evaporable water content in cement hydrated with 0, 10, and 30% silica fume and prepared at W/S ratios of 0.25 and 0.45.[10] In the figure W/C + SF = water/cement + silica fume ratio. The designation of all the mixes are shown within the figure. The specimens made at a higher W/S ratio (0.45) contain larger amounts of nonevaporable content. The sample with 0% silica fume attains a value of 19.2% at 28 days and 20.7% at 180 days. The corresponding values for samples containing 10% silica fume are 15.1 and 15.5% and for those containing 20% silica, 14.4 and 15.3%. Results for comparable mortars show similar trends; at 30% silica fume content the value increased from 11.25% at 14 days to 11.9% at 180 days. It is possible that in mortars the sand acts as a sink for well-crystallized $Ca(OH)_2$ and lowers the permeability of the paste phase of the mortar. At a W/S ratio of 0.25 further reduction in the nonevaporable water is evident. The lower amounts of nonevaporable water content in cements containing silica fume may be explained by the consumption of CH by silica fume with the formation of a C-S-H product containing less hydrated water than in that formed during normal hydration of cement.

Sellevold et al.[12] determined the nonevaporable water content in cement-silica fume

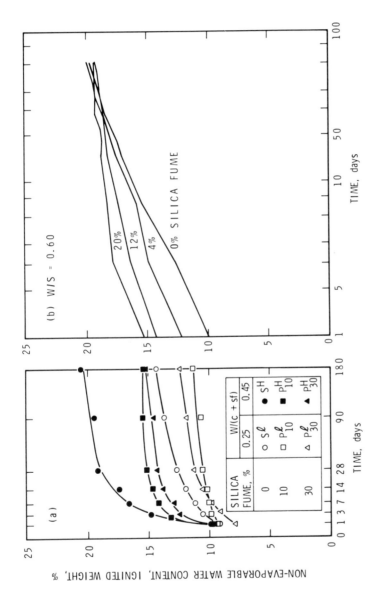

FIGURE 12. (a) Change in nonevaporable water content with time for various silica fume-cement mixtures. (b) Non-evaporable water content of cement hydrated with different amounts of silica fume.

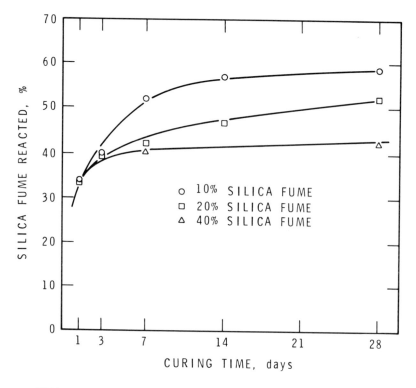

FIGURE 13.   Consumption of SiO₂ fume during the hydration of cement.

mixes (W/S = 0.6) hydrated with different amounts of silica fume (Figure 12b). An accelerating effect at early ages was observed with silica fume and $CaCO_3$ additions. It was concluded that these effects were inert filler effects. It has, however, been shown that $CaCO_3$ is not an inert filler in cement as it accelerates the hydration of the silicate phase and reacts with the hydration products.[3]

3. Consumption of SiO₂

The rate of consumption of silica fume in the cement-SiO₂ system provides some information on the hydration process. In Figure 13 the percentage of SiO_2 fume consumed at different times during the hydration of cement is plotted.[11] Within 1 day, about 30% of SiO_2 is consumed at all additions, viz., 10, 20, and 40%. At 28 days, 60, 52, and 43% SiO_2 has reacted in mixtures containing 10, 20, and 40% SiO_2, respectively. It can be concluded that silica fume not only reacts with Ca ions of the cement minerals but also with the CH derived from hydration. The hydration products were found to form around SiO_2 particles, indicating that the later reaction process is diffusion controlled.

C. Heat Evolution

Conduction calorimetry which measures the rate of heat development during the hydration of cement is another method of following the effect of silica fume on cement hydration. The rate of heat developed in cement pastes containing 0, 10, 20, and 30% silica fume is plotted in Figure 14.[10] The results are calculated on the basis of cement content to avoid dilution effects. The earlier peak below 1 hr is generally attributed to the aluminate reaction. The second broad hump occurring in the cement paste (reference) shows a peak effect at about 6 to 7 hr and this is caused by the C_3S hydration.

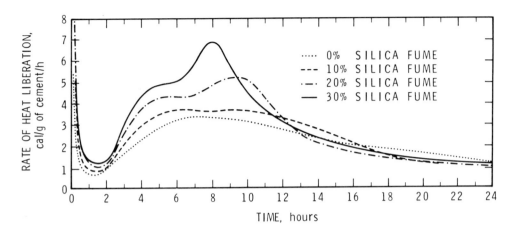

FIGURE 14.   Change in rate of heat evolution of C/S fume pastes with time.

Addition of silica fume changes the profile of the curves. An intense peak appears at about 5 hr and another sharp, intense peak at about 6 to 10 hr. The greater the intensities of these peaks, the larger the amount of added silica fume. The earlier peak can be associated with the C_3S hydration which occurs at an earlier time than in that containing no silica fume. This signifies acceleration of C_3S hydration and the other signifies the transformation of high sulfoaluminate hydrate to the low sulfoaluminate form.

The more prominent second peak occurring in the paste containing silica fume may be caused by its influence on the aluminate hydration. Another possibility is that this peak is due to the reaction of CH (formed by hydration) with silica fume. The intensity of this peak increases as the SiO_2 fume content is increased. The total amount of heat developed is increased when increasing amounts of silica fume are incorporated into cement. This shows that the overall reaction rate is increased (Figure 15).[10]

The total amount of heat developed during the hydration of cement depends on the initial surface area of cement. Kumar and Roy[13] studied the role of surface area of cement on heat developed during the hydration of pastes containing 0, 5, 10, and 15% silica fume (Figure 16). Cements of Blaine surface area 3750, 2320, 1950, and 1580 m²/ kg, designated C, C_1, C_2, and C_3, respectively, were used. The curves show that heat evolved per unit weight of solids for 10 to 15% silica fume mixture is lower compared to that for neat cement of any given fineness. If the heat evolution is compared on the basis of cement content the silica fume-containing cement shows higher heat evolution. It is also evident that a decrease of cement fineness for C through C_3S leads to a significant drop in the reactivity and total heat liberation during the early stages of hydration.

D. Surface Area

The surface area of cement paste hydrated in the presence of silica fume is affected (Figure 17).[11] At early times the surface area of the product containing silica fume is higher than that of the neat cement paste. This may partly be due to the silica fume itself having a high surface area of 23×10^3 m²/kg. In addition, at these periods the hydration is accelerated by silica fume. Within 7 days the surface area drops sharply. This is probably caused by the C-S-H product that is formed by the reaction of silica fume with $Ca(OH)_2$. Even at longer times the cement-silica fume pastes exhibit lower surface area. This has been explained by the densification of the microstructure which

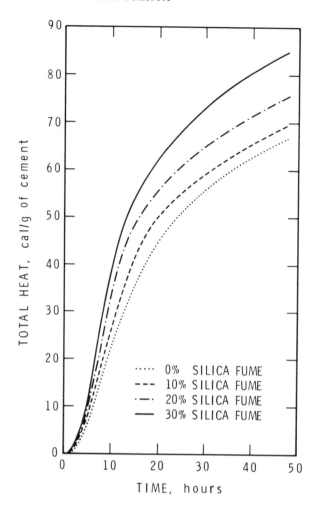

FIGURE 15.   Total heat evolved by C/S fume blends during hydration with time.

tends to decrease the N_2 penetration used for surface area determination. Another possibility is that at higher silica fume contents there is a reaction between the C-S-H formed by the Portland cement hydration and silica fume. This product may be responsible for decreased surface area.

E. CaO/SiO₂ Ratio of C-S-H

In the hydration of Portland cement the mean C/S ratio of C-S-H decreases with time from about 2.0 at 1 day to 1.4 to 1.6 after some years.[14] Because of the interaction of silica fume with CH and C-S-H, in the hydration of cement in the presence of silica fume, variation in the overall C/S of the product is expected. The C/S ratio would also depend on the amount of silica fume added.

It is generally reported that the hydration product formed in cements containing silica fume consists of C-S-H (I) type of product. The C/S values, varying between 0.9 and 1.3, have been reported using different techniques.[15-20] That a lower C/S product is formed in the presence of silica fume in cements is indicated by the thermograms of a cement paste containing 30% silica fume. In the thermograms of cement containing silica fume and hydrated for different times, an exothermic peak appears in all samples typical of the C-S-H (I) phase (Figure 18).[21]

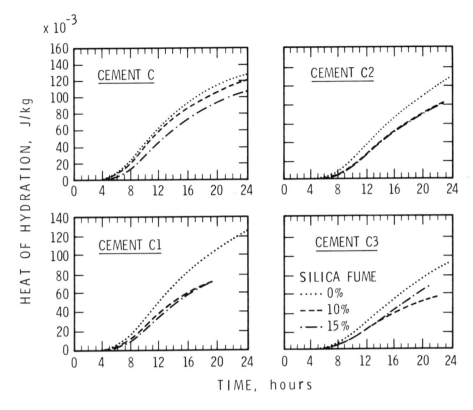

FIGURE 16.   Effect of cement fineness on the cumulative heat of hydration during first 24 hr of hydration for C/S fume blends.

F. Pore Solution Analysis

Analysis of the pore solution in Portland cement paste is one of the methods for investigating the mechanism of hydration and the possible long-term effects of Portland cement containing admixtures and additives. The importance of solution analysis on the mechanism of hydration of C_3S-SiO_2 pastes has already been described in Section II.C). In normal Portland cement pastes the pore solution consists essentially of alkali hydroxides with OH^- concentration as high as 700 mmol/l. In addition, other anions and cations are also present. It has been found that in the presence of ≤30% silica fume, enhanced early concentrations of alkalis and hydroxyl ions result, but after 1 day the effect is reversed, the concentrations of these ions getting progressively reduced.[22]

A systematic investigation of the concentration of Na^+, K^+, Ca^{2+}, OH^-, and SO_4^{2-} in cement hydrated with and without silica fume at different times has been carried out.[23] The data in Table 2 show the estimated amounts of anions and cations in cement hydrated for different periods and containing 0, 10, 20 and 30% silica fume. The results indicate that addition of silica fume causes substantial decrease in the concentrations of cations and anions.

In another study, Glasser and Marr[24] followed the Na^+ and K^+ concentrations in cement blended with 15% silica fume. At 180 days of hydration the reference cement contained 0.054 and 0.210 mol/l of Na^+ and K^+, respectively, and in the presence of silica fume the corresponding values were 0.02 and 0.058 mol/l. These reductions persisted even after the silica fume was physically consumed. Two mechanisms have been suggested to explain this observation. In normal cement pastes the C-S-H product

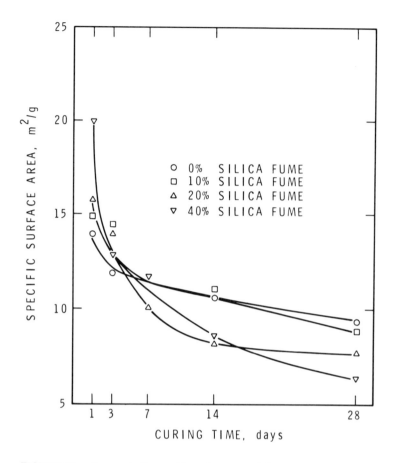

FIGURE 17.   Changes in the specific surface area of the cement paste containing silica fume.

has a C/S ratio of 1.4 to 1.6 and this composition has poor sorptive properties for alkalis.[25] The sorptive capacities for alkalis are improved as the C/S ratio in the C-S-H phase decreases. Thus, as the silica fume content increases, the possibility of the formation of lower C/S product is increased. It has also been suggested that the strength of sorption decreases in the order $Cs^+ > K^+ > Na^+$. According to another theory, in the presence of C-S-H with low C/S ratio, precipitation of alkali-containing compound can occur. These studies are useful to explain the role of silica fume in affecting the alkali-aggregate expansion reaction in concretes.

The influence of silica fume on the pH of the solution in the system cement-silica fume-water-chloride has been studied to assess the corrosion potential (Figure 19).[23] The plain cement paste shows a pH of about 13.9. Addition of silica fume results in a progressive decrease in pH. A corresponding reduction in the total available alkali metal cations in solution is also observed. However, even at an addition of 10 to 20% silica fume the pH value is in excess of that shown by a saturated $Ca(OH)_2$ solution (pH = 12.5). As the passivity of mild steel is maintained at pH >11.5, it would appear that at a dosage of 10 to 20% silica fume corrosion may not be a problem. There may, however, be indirect consequences of reduced alkalinity of concrete containing an excess of silica fume that needs further work.

The corrosion potential of steel in concrete is enhanced in the presence of chloride. Some studies have been carried out to estimate the amount of strongly bound chloride

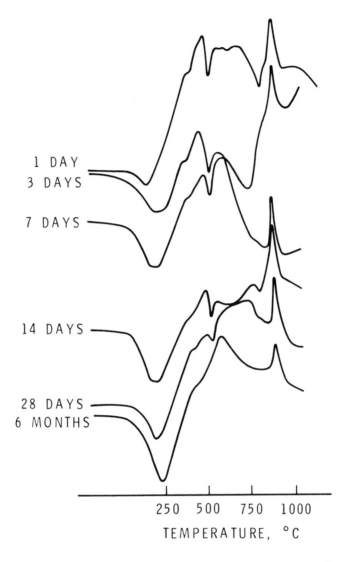

1 DAY
3 DAYS

7 DAYS

14 DAYS

28 DAYS
6 MONTHS

250   500   750   1000

TEMPERATURE, °C

FIGURE 18.   Differential thermograms of cement pastes hydrated to different times with 30% silica fume.

in C/S fume-chloride-H$_2$O systems. Figure 20 shows the concentration of chloride in C/S mixtures containing 0.4% chloride.[23] Incorporation of increasing amounts of silica fume leads to larger amounts of unbound Cl ions in the pastes. The dominating mechanisms by which the Cl ion is bound is by the formation of complex chloroaluminate.[26] The presence of silica fume results in the formation of decreased amounts of both Ca(OH)$_2$ and chloroaluminate. Lower amounts of chloroaluminate are formed because its solubility is increased as the pH of the solution is decreased. These results would suggest that reduced pH and increased amounts of chloride in Portland cement-silica blends would enchance corrosion in this system. It is also important to consider that the threshold concentration for corrosion in the presence of chloride increases steeply only above a pH of 11.75. In addition, these results cannot be applied to concrete containing silica fume because of other effects such as O$_2$ depletion, lower permeability, and limited mobility of Cl ions. Further work should be carried out to resolve these questions.

Table 2
ANALYSIS OF PORE SOLUTIONS IN CEMENT PASTES
CONTAINING SILICA FUME (W/S = 0.5)

Silica fume $\frac{}{(C + SF)} \times 100$	Curing time (days)	Ionic conc. (mmol/l)					Sum of conc. of detected ions (mg equiv./l)	
		Na⁺	K⁺	Ca²⁺	OH⁻	SO₄²⁻	Cations	Anions
0	7	263	613	1	788	23	876	834
	28	271	629	1	834	31	900	896
	56	332	695	3	839	44	1033	927
	84	323	639	2	743	27	966	797
10	7	161	388	1	486	20	551	526
	28	101	209	0	241	40	311	321
	56	117	218	1	233	30	337	293
	84	107	192	2	228	27	303	282
20	7	109	231	1	290	23	342	336
	28	59	109	1	91	33	170	157
	56	54	77	2	81	34	135	149
	84	51	69	2	78	25	124	128
30	7	75	143	1	152	33	220	218
	28	35	53	2	26	35	92	96
	56	40	39	5	11	36	83	89
	84	30	30	7	10	32	74	74

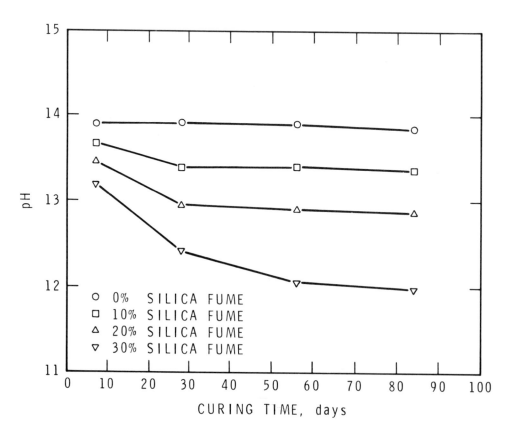

FIGURE 19.   Influence of silica fume content on pH values of cement pastes cured for different periods.

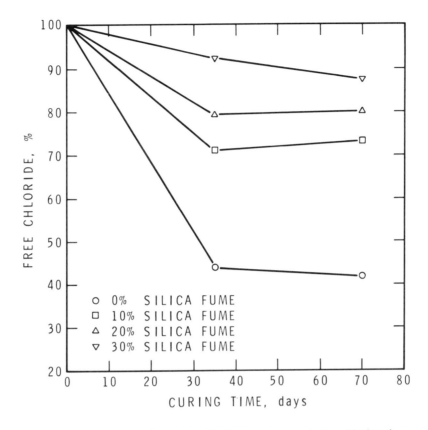

FIGURE 20. Percentage of total free chloride ions in pore solution of hydrated cement pastes containing silica fume.

G. Model

The mechanism of acceleration of C₃S in the presence of silica fume that was described earlier is also applicable to the hydration of Portland cement. The mechanism involves the changes in the concentration of Ca ions in the solution, the reaction of SiO₂ with Ca(OH)₂ and C-S-H, and nucleating effects.

A model for cement paste has been proposed by various workers and these have been discussed by Ramachandran et al.[1] A model for C/S fume paste has been described by Grutzeck et al.[17] and is different from that proposed for the cement paste. In this model silica fume comes into contact with water containing low concentrations of lime and absorbs moisture to form a silica-rich gel consuming most of the available water. The gel particles clump together in the interstices of the cement particles and agglomerate into large masses. In contrast to the network of radiating C-S-H crystals formed in normal cement pastes, in C/S pastes the anhydrous calcium silicate grains are surrounded by the development of silica gel within 15 min to 1 hr. At 3 hr CH in the pore solution of normal cement pastes forms discrete crystals, but in the presence of silica fume no such crystals of CH form. The excess lime reacts with the outer surface of the agglomerated gels to produce additional C-S-H which intergrows with the hydrating silicate grains. At 24 hr and later C-S-H development is enhanced in the silica fume pastes, and a matrix of C-S-H-cemented agglomerates of silica fume that are intergrown with C-S-H growing outward or inward from the calcium silicate grains is formed. This results in the formation of a rigid structure at earlier ages than that occurring in normal cement pastes. This model is based on density, compressive

Table 3

WATER REQUIREMENT AND POZZOLANIC ACTIVITY
INDEX OF VARIOUS TYPES OF SILICA FUMES

Silica fume	SiO_2%	Sp. surface area $(10^3 m^2/kg)$	Water requirement	Relative pozzolanic activity index
Si	94	20	138	100
Fe-Si (75%)	89	13	132	97
Fe-Si (75%) (heat recovery)	90	13	125	95
Fe-Si (50%)	83	15	106	56
Fe-Cr-Si	83	16	128	88
Ca-Si	53.7	—	104	80
Si-Mn	25	—	140	36

strength, and chemical analysis. This model is used to explain early strength, lower permeability, density, and thickening that are observed in cement pastes containing silica fume. Other possibilities such as porous structure, permeability, and interface effects have been used to explain these phenomena and are discussed in other chapters.

IV. SILICA FUME AS A POZZOLAN

A. C/S Fume-H₂O System

It has already been shown that in the hydration of cement containing silica fume the lime formed by the hydration process is consumed to different extents by silica fume additions (see Section III.B). The percentage $Ca(OH)_2$ consumed by silica fume has been used as an index of the pozzolanic activity of silica fume but the reported results show large variation. Traetteberg's data[18] would suggest that silica fume is only moderately pozzolanic. For example, at 10% silica fume addition only 2% CH was consumed during the hydration of cement at 14 days. According to Chatterji et al.,[27] who used the XRD method, the silica fume may exhibit poor pozzolanicity. In their experiments on cement-silica mixtures containing 17% silica, free CH was detected even after 2 years. Contrary to these reports, Buck and Burkes[28] reported high pozzolanic activity for the silica fume, as all CH formed during hydration at 28 days was consumed by silica fume. Others also came to similar conclusions.[10,11,19,29] The above investigations would suggest that not only differences in the type and amount of silica fume and W/S ratio, but also the techniques adopted for estimating the CH would explain the variability in reported results.

The pozzolanic activity index of a pozzolan may be assessed by ASTM C 311 and C 618 methods. In these methods the strength of the test mixture at 28 days should be at least 75% of the control. The maximum water requirement is 115% of the control and the mixture is cured for 1 day at room temperature and at 38°C for 27 days. This test cannot be applicable to C/S fume mixture for the following reasons. The water requirement is more than the specified limit (Table 3).[30] For strength development, curing temperature need not be as high as 38°C. In addition, in comparison with other pozzolans, generally smaller amounts of silica fume are added to Portland cement.

Table 3 indicates that the water requirement depends on the SiO_2 content and surface area; the pozzolanic activity shows a wide variation, depending on these two factors. Generally, the richer the SiO_2 content, the higher the pozzolanic activity.

The pozzolanic activity, based on chemical methods, may not yield a good correlation with that based on strength measurements. This is because the strength may also

Table 4
STRENGTH DEVELOPMENT IN
Ca(OH)$_2$-POZZOLAN MIXTURES

Silica fume:Ca(OH)$_2$ ratio	Flow	W/S ratio	Compressive strength (MPa) (days)			
			7	28	90	180
2:1	93	1.06	8.8	11.5	11.3	12.7
1:1	95	0.98	9.3	11.8	11.6	10.9
1:2.25	101	0.93	5.5	7.1	6.9	7.1

increase by changes in the pore structure and interfacial regions caused by silica fume addition. It is also suggested that the decrease in the amount of CH crystals (which are a source of weakness in normal cement pastes), rather than the increase in C-S-H or reduction in porosity, is responsible for higher strengths in the cement paste containing silica fume.[31]

The high pozzolanic activity of silica fume allows it to be used in blended cements containing other pozzolans. The early strengths developed in normal pozzolan-cement mixes are low and this can be improved by adding silica fume to these pozzolans.[32] For example, the strength of fly ash-containing concrete is lower than that of control concrete (without fly ash) at early ages. However, by partial replacement of fly ash with silica fume the strength at 3 days has been found to be similar to that of control concrete[32] and the free lime contents in various mixes were consistent with the strength data.

B. Ca(OH)$_2$-Silica Fume-H$_2$O System

As the system Ca(OH)$_2$-silica fume-H$_2$O is simpler to study than that containing Portland cement, some attention has been directed to this system for assessing the pozzolanicity of silica fume.

1. Pozzolanic Activity Index

The pozzolanic activity index of a pozzolan with respect to reaction with lime is covered under ASTM Standards C 311 and C 618. These tests specify the minimum strength requirements. The mixture is stored at 55 ± 1.7°C for 6 days after it is initially cured at 1 day at about 23°C. The minimum strength at 7 days should be 5.5 MPa. This test cannot be applied directly to silica fume because its water requirement may be higher and the mixture develops strengths at a faster rate than other pozzolans, even at the ambient temperatures. In Table 4, the strength development in Ca(OH)$_2$-silica fume mixtures containing different amounts of silica fume is compared. Maximum strengths are attained within 28 days. The optimum ratio is 1:1 for strength development.[28] The results confirm that silica fume is an effective pozzolan.

2. Reaction Rate and Products

Conduction calorimetric investigations in the Ca(OH)$_2$-SiO$_2$-H$_2$O systems have demonstrated that as soon as water comes into contact with Ca(OH)$_2$-SiO$_2$ mixtures a heat peak is developed. This heat effect cannot be explained by the combined heat developed by the dissolution of pure CH and heat of wetting of silica fume. The heat effect is therefore mainly attributed to the reaction between Ca(OH)$_2$ and silica fume occurring within a few minutes. No other peak is observed at later times.

The rate of disappearance of Ca(OH)$_2$ from the Ca(OH)$_2$-SiO$_2$ fume mixtures con-

Table 5

REACTION OF Ca(OH)₂ IN Ca(OH)₂-SiO₂ MIXTURES

	Hours					Days				
Time	0	1	3	5	12	1	3	7	14	31
Ca(OH)₂% (with SiO₂, 20 × 10³m²/kg)	55	—	—	53.9	—	50	40.2	27.7	12.3	3.5
Ca(OH)₂% (with SiO₂, 90 × 10³m²/kg)	55	27.3	23.5	20	4.1	1.4	0	0	0	0

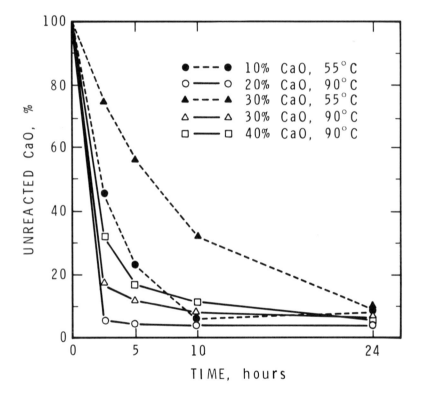

FIGURE 21.   The effect of temperature on the reaction between Ca(OH)₂ and silica fume.

taining SiO_2 of surface area 20×10^3 m²/kg or 90×10^3 m²/kg is compared in Table 5.[6] These results indicate that the reaction rate depends on the surface area of SiO_2. Lime was consumed completely within 1 day when high surface area SiO_2 was used.

At higher temperatures (using steam) molded products may be produced using Ca(OH)₂-silica fume mixtures. The rate of reaction is faster than that occurring at normal temperatures. The rate of reaction in this system carried out at 55 or 90°C can be judged from Figure 21.[33] At 90°C in 2 to 3 hr as much as 68 to 95% of added CaO has reacted. At the same time only 25 to 55% of CaO has reacted at 55°C. At 24 hr most CH has been consumed by the reaction.

The reaction product formed in the Ca(OH)₂-silica fume mixtures at ordinary temperatures consists of CSH(I). Buck and Burkes[28] found that XRD generally exhibited peaks that could be attributed to CSH(I) (Figure 22). Even if the original mixture

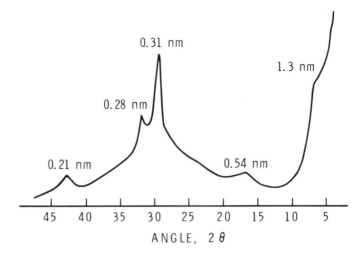

FIGURE 22.   XRD chart of the product formed in the $Ca(OH)_2$-SiO_2 system.

contained higher CaO (e.g., $CaO/SiO_2 = 1.7$), no CSH(II) formed. Wu and Young[6] reported that the C-S-H product formed at various periods of curing had a C/S ratio in the range 1.1 to 1.26. These values were slightly lower than those formed in the C_3S pastes. Using a W/S = 1500, Grutzeck et al.[16] found small amounts of C-S-H in the $Ca(OH)_2$ solution-SiO_2 fume system at 21°C, but at 38°C the product contained CSH(I). At higher temperatures of curing, other products (crystalline) are expected to form. At a curing temperature of 80°C, formation of jennite has been reported.[34] Xonotlite and CSH(I) were detected in the mixtures cured at 190 to 200°C.[33]

3. Mechanism

Grutzeck and co-workers have developed a mechanism for the pozzolanic reaction of silica fume with $Ca(OH)_2$ in aqueous solutions. It is based on the analysis of ions in solution, and application of other methods such as IR, X-ray, and SEM.[16] According to this mechanism, in contact water, silica goes into solution and within 1 hr or less the silica in solution forms an amorphous silica-rich, Ca-poor gel on the surface of silica fume particles. The silica fume particles also form agglomerates. With time, this silica-rich, Ca-poor coating dissolves and the agglomerates of silica fume start reacting with $Ca(OH)_2$ to form C-S-H. The initial reaction described above is surface dependent. The formation of C-S-H by the pozzolanic reaction is temperature dependent. This model has been extended to Portland cement systems, for explaining the higher water demand and premature hardening effects (see Section III.G).

REFERENCES

1. **Ramachandran, V. S., Feldman, R. F., and Beaudoin, J. J.,** *Concrete Science,* Heyden & Son, London, 1981.
2. **Stein, H. N. and Stevels, J. M.,** Influence of silica on $3CaO.SiO_2$, *J. Appl. Chem.,* 14, 338, 1964.
3. **Ramachandran, V. S.,** Admixture and additive interactions in the cement-water system, 83, 13, 1986.
4. **Kurdowski, W. and Nocun-Wczelik, W.,** The tricalcium silicate hydration in the presence of active silica, *Cement Concrete Res.,* 13, 341, 1983.

5. Ogawa, K., Uchikawa, H., Takemoto, K., and Yasui, I., The mechanism of the hydration in the system C₃S-pozzolana, *Cement Concrete Res.,* 10, 683, 1980.
6. Wu, Z. Q. and Young, J. F., The hydration of tricalcium silicate in the presence of colloidal silica, *J. Mater. Sci.,* 19, 3477, 1984.
7. Tamas, F. D., Sarkar, A. K., and Roy, D. M., in *Hydraulic Cement Pastes: Their Structure and Properties,* Cement and Concrete Association, Slough, England, 1976, 55.
8. Roy, D. M., Skalny, J., and Diamond, S., Effects of blended materials on the rheology of cement pastes and concretes, in Concrete Rheology, Proc. Symp. Annu. Meet. Mater. Res. Soc., Boston, 1982, 152.
9. Bache, H. H., Densified cement/ultra-fine particle-based material, Presented at 2nd Int. Conf. Superplasticizers in Concrete, June 10-12, 1981, Ottawa.
10. Cheng-yi, H. and Feldman, R. F., Hydration reactions in Portland cement-silica fume blends, *Cement Concrete Res.,* 15, 585, 1985.
11. Ono, K., Asaga, K., and Daimon, M., Private communication.
12. Sellevold, E. J., Bager, D. H., and Jensen, E. K., Microsilica-cement pastes: hydration and pore structure, presented at the Int. Conf. on the Use of Fly Ash, Silica fume, Slag and Other Mineral By-Products in Concrete, Montebello, Canada, August 1983.
13. Kumar, A. and Roy, D. M., A study of silica fume-modified cements of varied fineness, *J. Am. Ceram. Soc.,* 67, 61, 1984.
14. Taylor, H. F. W., Portland cement-hydration products, in *Instructional Modules in Cement Science,* Roy, D. M., Ed., Materials Education Council, 1985.
15. Meland, I., Influence of condensed silica fume and fly ash on the heat evolution in cement pastes, Proc. 1st Conf. on the Use of Fly Ash, Silica Fume, Slag and Other Mineral By-Products in Concrete, Montebello, Canada, Vol. 2, Malhotra, V. M., Ed., 1983, 665.
16. Grutzeck, M. W., Atkinson, S., and Roy, D. M., Mechanism of hydration of condensed silica fume in calcium hydroxide solutions, *ibid.,* p. 643.
17. Grutzeck, M. W., Roy, D. M., and Wolfe-Confer, D., Mechanism of hydration of Portland cement composites containing ferrosilicon dust, Proc. 4th Int. Conf. Cement Microscopy, Las Vegas, 1982, 193.
18. Traetteberg, A., Silica fumes as a pozzolanic material, *Il Cemento,* 75, 369, 1978.
19. Sellevold, E. J., Bager, D. H., Jensen, E. K., and Knudsen, T., Silica fume-cement pastes: hydration and pore structure, in *Condensed Silica Fume in Concrete,* Institute of Building Materials, Norwegian Institute of Technology, Trondheim, 1982, 20.
20. Hjorth, L., Silica fumes as additions to concrete, in *Characterization and Performance Prediction of Cement and Concrete,* Engineering Foundation, Henniker, 1982, 165.
21. Meland, I., Hydration of Blended Cements, Nordic Concrete Research publ. no. 2, 1983, 183.
22. Diamond, S., Effects of microsilica (silica fume) on pore-solution chemistry of cement pastes, *J. Am. Ceram. Soc.,* 66, 82, 1983.
23. Page, C. L. and Vennesland, O., Pore Solution Composition and Chloride Binding Capacity of Silica Fume-Cement Pastes, Materials and Construction publ. no. 91, 1983, 19.
24. Glasser, F. P. and Marr, J., The effect of mineral additives on the composition of cement pore fluids, Br. Ceramics Proc., Stoke-on-Trent, England, 1984, 419.
25. Glasser, F. P., Rahman, A. A., Crawford, R. W., McCulloch, C. E., and Angus, M. J., in *Testing, Evaluation and Shallow Land Burial of Low and Medium Radioactive Waste Forms,* Krischev, W. and Simon, R., Eds., Hardwood Academic, London, 1984.
26. Ramachandran, V. S., *Calcium Chloride in Concrete,* Applied Science, London, 1979.
27. Chatterji, S., Thaulow, N., and Christensen, P., Pozzolanic activity of by product silica-fume from ferro-silicon production, *Cement Concrete Res.,* 12, 781, 1982.
28. Buck, A. D., and Burkes, J. P., Characterization and reactivity of silica fume, Proc. 3rd Int. Conf. on Cement Microscopy, Houston, March 1981, 279.
29. Nelson, J. A. and Young, J. F., Additions of colloidal silicas and silicates to Portland cement pastes, *Cement Concrete Res.,* 7, 277, 1977.
30. Aitcin, P. C., Pinsonneault, P., and Roy, D. M., Physical and chemical characterization of condensed silica fumes, *Am. Ceram. Soc. Bull.,* 63, 1487, 1984.
31. Scriverer, K. L., Boldie, K. D., Halse, Y., and Pratt, P. C., Characterization of microstructure as a systematic approach to high strength cements, in *High Strength Cement Based Materials,* Vol. 42, Young, J. F., Ed., Materials Research Society, 1984, 39.
32. Mehta, P. K. and Gjorv, O. E., Properties of Portland cement concrete containing fly ash and condensed silica fume, *Cement Concrete Res.,* 12, 587, 1982.
33. Kurbus, B., Bukula, F., and Gabrovsek, R., Reactivity of SiO₂ fume from ferrosilicon production with Ca(OH)₂ under hydrothermal conditions, *Cement Concrete Res.,* 15, 134, 1985.
34. Hara, N. and Inoue, N., Formation of jennite from fumed silica, *Cement Concrete Res.,* 10, 677, 1980.

Chapter 5

PORTLAND CEMENT-SILICA FUME BLEND PASTES

I. INTRODUCTION

Various studies have shown that the addition of silica fume, even in moderate quantities, can have a very significant effect on the course of the hydration of Portland cement.[1-5] It is to be expected, therefore, that silica fume will also have a major effect on the physical properties of normal hydrated cement paste. This chapter discusses the effect of silica fume on the cement pastes in terms of porosity, density, permeability and pore-size distribution, their interrelationships with some mechanical properties, and frost resistance.

Published data on physical properties of Portland cement-silica fume blend pastes are sparse. Most of the data are taken from results for blends at 10 and 30% silica fume content.[5-7] The silica fume used was reactive in nature, containing 95% silica, and having a surface area of 21 m²/g.

II. MICROSTRUCTURE

Figure 1 shows a scanning electron micrograph of silica fume. Generally, it appears as a round particle of average size, 0.2 to 0.3 μm. Some of these particles of less than 0.1 μm are attached on the surfaces of large particles.

Figures 2A, B, and C are representatives of micrographs of cement pastes hydrated at a W/S ratio of 0.45 for 90 days and containing 0, 10 and 30% silica fume. The major difference is, whereas cement paste without silica fume is more porous in appearance and the presence of $Ca(OH)_2$ is very apparent, the blended pastes appear more consolidated and have an almost vitreous-like appearance, although the total porosity of blends is not too dissimilar from the reference material.[3] This vitreous appearance may be due to the higher proportion of C-S-H and differences in the mode of fracture during preparation of the surfaces. The paste containing no silica fume shows some columnar clusters due to $Ca(OH)_2$. At 90 days, almost all $Ca(OH)_2$ in the blended pastes must have reacted with the silica fume and, hence, it cannot be discerned.

The micrographs of paste hydrated for 1 day at a W/(C + SF) of 0.25 are shown in Figures 3A, B, and C. The blend containing 30% silica fume has a much more consolidated structure than others. Although $Ca(OH)_2$ was detected by the microprobe in the sample containing 0 or 10% silica fume, there was absence of lime in that containing 30% silica fume. It was difficult, by the microprobe and scanning microscope, to accurately determine the CaO/SiO_2 ratio of the C-S-H product due to interference effects from SiO_2 and CH. Work on mortars has indicated a C/S (cement/silica) ratio of 0.9 to 1.3.[8-10]

III. PORE STRUCTURE

A. Measurement Techniques

Three types of techniques may be used to determine the total porosity.

1. Helium Comparison Pycnometry

This technique has been used with some success for the determination of true density

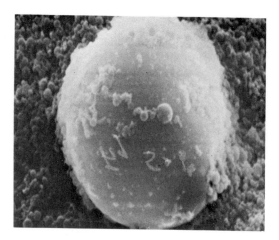

FIGURE 1.   Electron micrograph of unreacted silica fume.

and porosity of hydrated cement.[11] The instrument consists of two cylinders (one is a reference cylinder and the other contains the sample) with pistons into which He is filled at a pressure of 1 atm. After isolating the pistons, the gas is compressed to 2 atm by moving both pistons. This results in the decrease of volume by 50% and pressure is doubled. The sample piston and the reference piston are moved with respect to the differential pressure indicator by keeping the pressure in the two cylinders the same. Using the gas laws and assuming an ideal gas, absolute volume of a solid can be determined. If the apparent volume is known, the total porosity of the solid can be calculated.

2. Mercury Porosimetry

The principle of the Hg porosimeter consists of determining the quantity of Hg that is forced into the pores of the materials at different pressures. The minimum pore diameter penetrated by Hg depends on the applied pressure. For example, at 5 psi pores of 35 μm diameter are penetrated, whereas at 60,000 lb/in.^{2} (414 MPa), pores down to 30Å diameter are intruded. Mercury porosimeter is used to determine pore sizes in the range 25Å to 1000 μm.

3. Porosity Using Water as a Fluid

In this technique, damp dry specimens are placed in desiccators conditioned with salt solutions to 11% relative humidity. The weight change from the damp dry condition is recorded as a percent of the volume of the specimen in the d-dried state (a method of drying of a sample by bringing it to equilibrium at a water vapor pressure of 5×10^{-4} mmHg; this method is used to expel all the physically held water in the cement paste but some combined water is also lost). The total water held from the damp-dry to the d-dry condition is also used as a measure of the total porosity.[12]

B. Porosity Measurements: Their Relevance

1. Mercury Porosimetry and Helium Pycnometry

Porosity measurements have been made on pastes containing 0, 10, and 30% silica fume and prepared at W/(C + SF) of 0.25 and 0.45 for up to 180 days hydration.[6] Some results obtained from mercury porosimetry are shown in Figure 4; the symbols used for the specimens are described in Table 1. At both W/(C + SF) ratios the lowest porosity occurs with the 30% silica fume blend and is 10% at a W/(C + SF) of 0.25

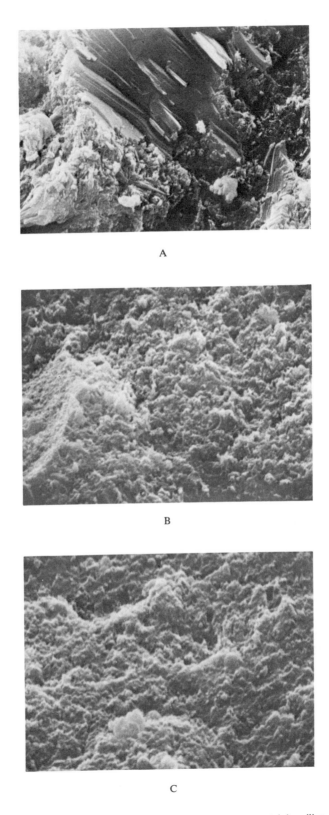

A

B

C

FIGURE 2.   Electron micrographs of cement pastes containing silica fume hydrated for 90 days at a W/(C + SF) = 0.45. (A) 0% silica fume, (B) 10% silica fume, (C) 30% silica fume.

A

B

C

FIGURE 3.   Electron micrographs of cement pastes containing silica
fume hydrated for 1 day at a W/(C + SF) = 0.25. (A) 0% silica fume,
(B) 10% silica fume, (C) 30% silica fume.

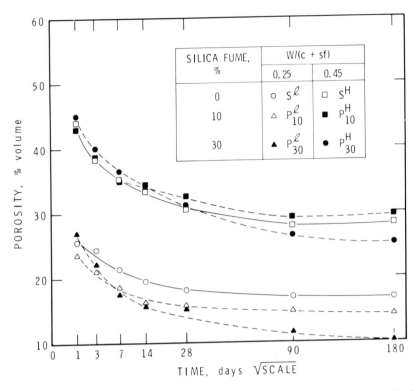

FIGURE 4.   Porosity of various cement pastes measured by mercury porosimetry vs. time of hydration.

Table 1
NOMENCLATURE FOR
MIXES

Silica fume (%)	W/(C + SF)[a]	
	0.25	0.45
0	S^l	S^H
10	P^l_{10}	P^H_{10}
30	P^l_{30}	P^H_{30}

[a]   Water/cement plus silica fume
ratio (by weight).

and 25.2% at a W/(C + SF) of 0.45. The 10% silica fume blend at a W/(C + SF) of 0.45 has the highest porosity.

Porosity measurements by helium pycnometry and those by Hg porosimetry do not differ significantly for the specimens at a W/(C + SF) of 0.25. However, a W/(C + SF) of 0.45, significant differences occur (Figure 5). For specimen S^H, the porosity values are very similar, remaining close to the line of equality, but values for P^H_{10} deviate after 7 days and after 14 days for P^H_{30} the mercury intrusion technique, giving higher porosity values.

2. Porosity Using Water as a Fluid

The weight of water lost between damp-dry and 11% relative humidity condition as

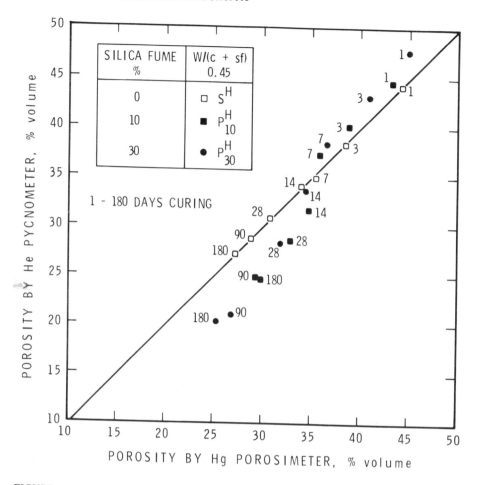

FIGURE 5.   Comparison of porosities determined by helium pycnometry and mercury porosimetry on specimens made at a W/(C + SF) of 0.45.

a percent of the volume of the oven-dried specimens has been measured for the various preparations.[6] The ratio of these data with the porosity determined by helium pycnometry is plotted as a function of curing time (Figure 6). After 1 day, most of the points fall between ratios of 0.86 and 0.93, and at 3 days, between 0.93 and 1.02; after 3 days, ratios for P_{10}^i and P_{30}^i increase rapidly, but S^i and S^H are scattered near 1.10. From 14 to 180 days, ratios for P_{10}^i, P_{30}^i and P_{30}^H are in excess of 1.50, that for P_{10}^H near 1.45 and those for S^i and S^H near 1.10. This indicates that following 3 days of curing most blends no longer absorb helium in all of their pores.

Helium pycnometry and mercury porosimetry methods are known to give relatively accurate porosity results for plain cement paste at a W/C (water/cement ratio) above 0.40.[13] Results using water as fluid, however, may not be accurate because it involves selection of the right fluid density as hydration proceeds and the size of the pores diminishes. The other problem is that decomposition of hydrates occurs during drying of specimens from 100 to 11% relative humidity. The results presented in Figure 6 for plain pastes are used for calibration purposes. It is assumed that the same correlation would apply for blended pastes. Consequently, all results obtained with water as a fluid for 1-day curing were divided by 0.90, those for 3 and 7 days by 1.00, and those for 14 days and over by 1.10 to obtain the porosity of the blends.

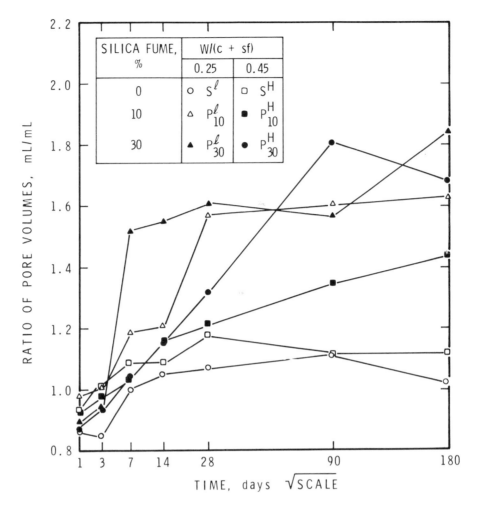

FIGURE 6.   Ratio of pore volumes (unadjusted value) by water technique divided by value by helium pycnometry vs. time.

A plot of adjusted porosities (determined by this water technique for all the specimens cured from 1 to 180 days) versus porosity determined by mercury porosimetry is compared in Figure 7. Points for specimens S^l, S^H, and P^H_{10} are scattered fairly closely round the line of equality, but points for P^l_{10}, P^l_{30}, and P^H_{30} deviate widely, displaying higher values by the water technique, especially after longer curing periods. It is clear that mercury intrusion does not measure the correct porosity under these conditions. Other results have also indicated that volume based on evaporable water contents is significantly higher than mercury intrusion volume.[12]

Helium pycnometry and mercury porosimetry give similar porosities for pastes above W/C of 0.40, and densities obtained from these results are consistent with those of known C-S-H material.[11,13] In fly-ash and slag blends mercury intrusion gives higher porosities than helium pycnometry or methanol saturation owing to the presence of isolated pores into which mercury enters as a result of partial breaking of the structure.[14] This seems to occur for silica fume blends at curing times as early as 7 days.

Using water loss between the damp-dry condition and 11% relative humidity, calibrating this technique with plain pastes and helium pycnometry, and comparing the

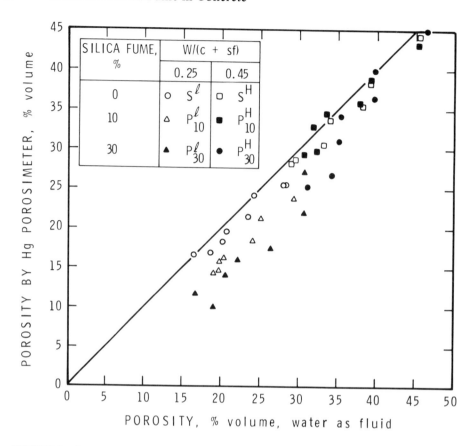

FIGURE 7.   Porosity determined by mercury porosimetry vs. adjusted porosity values determined by water technique.

results with those obtained by Hg has shown that mercury does not reach all the pores of specimens P^l_{10}, P^l_{30}, and P^H_{30}, although it appears to reach most of the pores of S^l, S^H, and P^H_{10}. Calibration of this technique (drying to 11% relative humidity) incorporates the density values of water in different sized pores and avoids inclusion of the water from decomposing hydrates as pore water during drying below 11% relative humidity.

C. Pore-Size Distribution

Pore-size distribution curves for the W/(C + SF) = 0.25 mixes with 0, 10, and 30% silica fume are shown in Figure 8. These were determined by Hg porosimetry.[6] Total porosities decrease with time of hydration. At 180 days they are 16.81, 14.43, and 10.08% with 0, 10, and 30% silica fume, respectively. The curves are of decreasing slope or are essentially concave to the pore-size axis for 0% silica fume (S^l) becoming increasingly convex with increasing silica fume content and longer curing times. The threshold pore diameter (diameter at which the slope of the volume-diameter curve increases abruptly) of the curves decreases with increase in silica fume content. At pore diameter greater than the threshold diameter the pore volume spread between 1 and 180 days increased with silica fume content. With 30% silica fume content, P^l_{30}, specimens hydrated for 28 to 180 days had higher pore volumes (for pores down to about 20 nm) than those hydrated for 7 days.

The pore-size distribution curves for W/(C + SF) = 0.45 mixes are shown in Figure 9 for 0, 10, and 30% silica fume. Total porosities of 28.30, 29.80, and 25.3% were

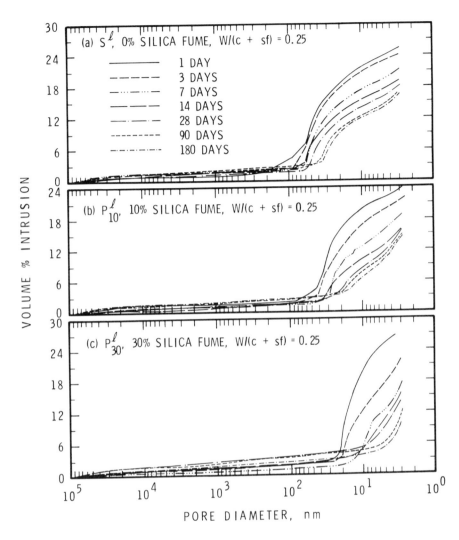

FIGURE 8. Pore-size distribution curves of cement pastes with different silica fume contents (W/(C + SF) = 0.25). (a) 0% silica fume, (b) 10% silica fume, (c) 30% silica fume.

observed for 0, 10, and 30% silica fume content at 180 days. Threshold diameters decrease with silica fume content, and spread of pore volume between 1 and 180 days also increases with silica fume content. This trend was also observed for the W/(C + SF) ratio of 0.25. In the range of pore diameters up to 100 nm and above, the pore volume for specimen P_{30}^H is largest after 180 days of hydration, being over 4% compared with values for short periods of hydration (e.g., 2.5% after 7 days).

The curves change from concave (to the pore-size axis) for 0% (S^H) to increasingly convex with increased curing time. For specimens containing silica fume, the slopes of the curves at maximum intrusion pressure increase with time, attaining maximum values between 14 and 90 days. This factor (slope) is highest where the $Ca(OH)_2$ content of the specimen is low as shown in Figure 10; those without silica fume where the $Ca(OH)_2$ content is high are relatively low and constant in slope. The convex shape of the curve at high pressures is due to the breaking of the pore structure by mercury at these high pressures and mercury entering large but relatively isolated pores. The extent of the isolation of pores appears to be related to the $Ca(OH)_2$ content. Both Hg intru-

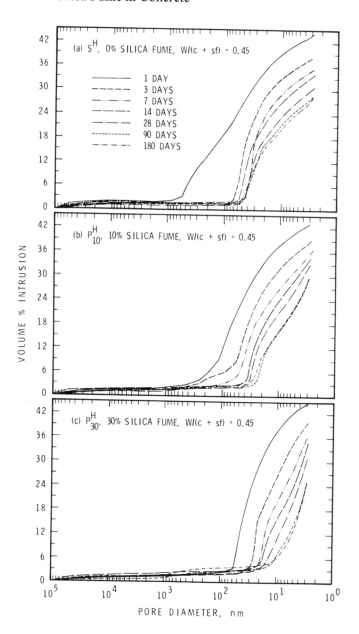

FIGURE 9.   Pore-size distribution curves of cement pastes with different silica fume contents (W/(C + SF) = 0.45). (a) 0% silica fume, (b) 10% silica fume, (c) 30% silica fume.

sion and $Ca(OH)_2$ have been measured for several fly-ash and slag cement pastes. Slopes of the pore distribution curves are calculated and shown in Figure 11. Where the $Ca(OH)_2$ content is low or decreasing, the slope of the distribution curve is high or increasing and vice versa. It is evident from consideration of the slopes alone in Figures 10 and 11 that silica fume reacts starting from curing times earlier than 3 days, while in two fly ashes (marginally type F) reaction commences only between 28 and 90 days. On the other hand, the slag content has a low $Ca(OH)_2$ content at all times and has a high slope before 3 days. It has been concluded that the $Ca(OH)_2$-CSH interface and

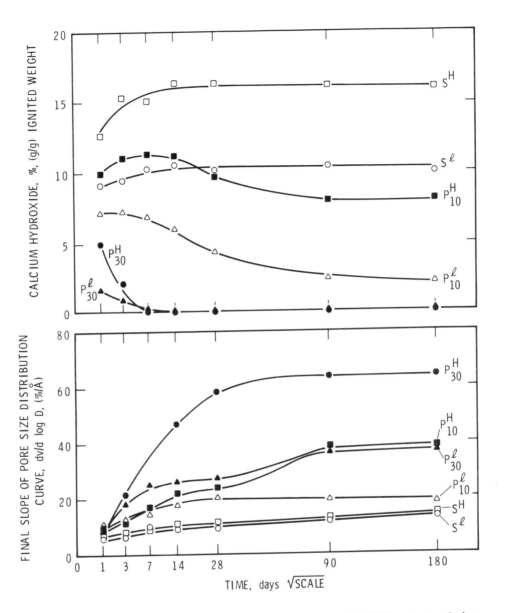

FIGURE 10. Slope of mercury intrusion curve at maximum pressure and Ca(OH)₂ content vs. hydration time for silica fume blends.

the Ca(OH)₂ crystals themselves provides the pathway for migration of Hg through the pores.[14] When the Ca(OH)₂ content decreases and Ca(OH)₂ interfaces are replaced by CSH-CSH interfaces, permeability of the body declines. Access to many of the pores by Hg during Hg intrusion experiments is only possible by rupture of the structure. Some pores are not accessible at all by Hg. These conclusions are drawn from Hg re-intrusion experiments.[6,16]

A comparison of pore-size distribution for various blended pastes prepared at high W/C ratio (W/C = 0.75) and hydrated at 7, 28, and 90 days is shown in Figures 12, 13, and 14, respectively.[17] Threshold diameters decrease with silica-fume addition even after only 7 days hydration. After 90 days hydration, threshold diameters decrease as: Portland cement > Portland cement + fly-ash Portland cement + fly-ash + silica fume

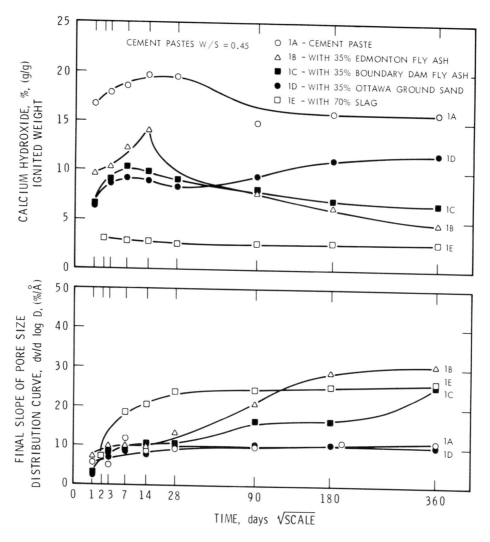

FIGURE 11.   Slope of mercury intrusion curve at maximum pressure and Ca(OH)$_2$ content vs. hydration time for fly ashes and slag.

> Portland cement + silica fume. Slopes of these curves at maximum intrusion pressure increase from Portland cement to the silica fume blend in the reverse order as for threshold diameters and the Ca(OH)$_2$ contents are lowest for the greatest slope.

D. Pore Discontinuity

The Hg intrusion technique, in some cases, permits determination of porosity. It is possible the initial Hg intrusion causes damage to the pore structure of the specimen and, hence, the pore volume-pore size plot may not be realistic. This can be verified by removing the Hg from the pores and repeating the Hg intrusion experiment.[16]

Specimens of pastes, S^H, P^H_{10}, and P^H_{30}, hydrated for 90 days and used for Hg intrusion experiments were prepared for second intrusion by heating at 105°C for 6 to 10 weeks in vacuum.[6] Since samples have been treated at this temperature prior to first intrusion, it is not felt that this treatment will create artefacts. The final weight of the specimens indicated that most of the Hg has been removed. Results are presented in Figure 15 for S^H, P^H_{10}, and P^H_{30}. Results for specimen S^H show little difference between first and second

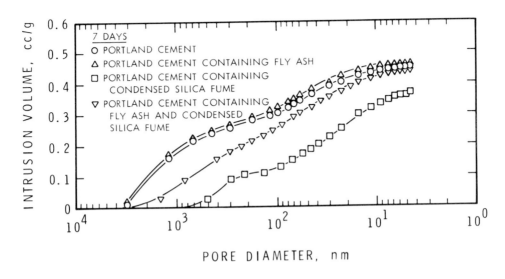

FIGURE 12. Pore-size distribution curves of cement paste with different blends, 7 days hydrated (W/ S = 0.75).

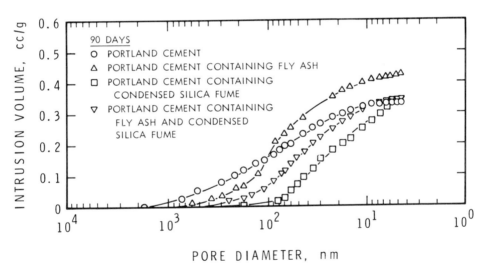

FIGURE 13. Pore-size distribution curves of cement paste with different blends, 28 days hydrated (W/ S = 0.75).

intrusion, although the threshold pore diameter is slightly larger for the second intrusion. The difference increases with silica fume content, the pore volume becoming greater in the pore diameter range 3×10^1 to 3×10^4 nm for the second intrusion. The curve for P^H_{10} changes character, becoming concave to the pore diameter axis. In general, a higher proportion of the pore volume is intruded at higher pore sizes.

These results indicate that relatively discontinuous pores are formed in pastes of silica fume blends and that these pores are of significantly larger diameter than are indicated by initial Hg intrusion experiments. The consequence of this is that permeabilities of these bodies can be greatly affected by pore discontinuity.

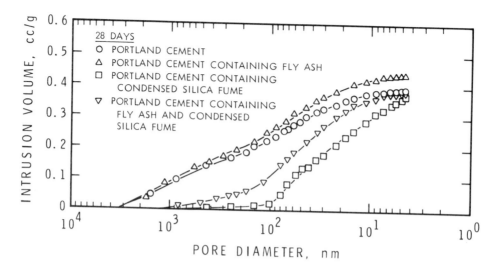

FIGURE 14.   Pore-size distribution curves of cement paste with different blends, 90 days hydrated (W/S = 0.75).

E. Density of Hydrate Products

The implications of measuring porosity using water as a fluid are described under Section III.B. Densities calculated from the adjusted water porosities for all the specimens as a function of curing time are presented in Figure 16. Values for the blends do not change significantly beyond 28 days. P^{H}_{30} and P^{H}_{10} have values of approximately 2.05 and 2.14 g/mℓ, respectively, after 180 days curing; S^{I} and S^{H} have values of 2.40 and 2.22 g/mℓ, respectively. Values for the plain pastes by He pycnometry are 2.39 and 2.24 g/mℓ, respectively, and those by Hg porosimetry are 2.35 and 2.23 g/mℓ, respectively, at 180 days. There is thus good correspondence. Values for the blends using He pycnometry and Hg porosimetry are much lower: 1.78 and 2.00 g/mℓ for P^{H}_{30} and P^{H}_{10}, respectively, with He; and 1.83 g/mℓ for P^{H}_{30} and 2.10 g/mℓ for P^{H}_{10} with Hg. These values (both for plain pastes and for blends) are for the d-dried material. Published data of densities of various Ca silicates show that it is unlikely for these products to have densities significantly below 2.00 g/mℓ.[15]

F. Surface Area

The surface area of the hydrating blends is dependent on several factors. One of the factors is the surface area of silica fume itself which is about 21 m²/g measured by nitrogen adsorption. However, the W/(C + SF) and the method of drying are also significant for the final value measured by nitrogen adsorption since both the accessibility of the nitrogen molecule to all surfaces and aging due to drying have an influence.

Surface area values have been obtained for pastes containing 10 and 30% silica fume and made at a W/(C + SF) of 0.25 and 0.45. Nitrogen adsorption method was used.[6] At a W/(C + SF) of 0.25, the specimen without silica fume had a value that remained constant (≈10 m²/g) from 7 to 180 days of curing, while with 10 or 30% silica fume the values varied from 7 to about 9 m²/g. At a W/(C + SF) of 0.45 all values varied from about 20 to 22 m²/g after curing from 7 to 180 days.

IV. PERMEABILITY AND DIFFUSION

Only a limited amount of work has been carried out on the permeability of silica-

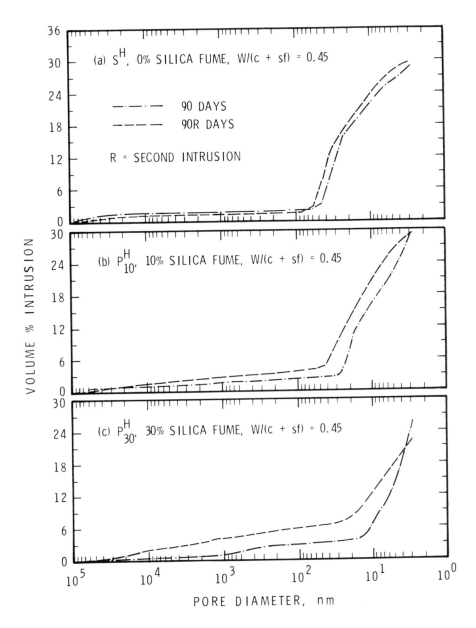

FIGURE 15. Pore-size distribution curves of cement paste with different silica fume contents, first and second intrustion (W/(C + SF) = 0.45, 90 days curing). (a) 0% silica fume, (b) 10% silica fume, (c) 30% silica fume.

fume-Portland cement pastes. Permeability to water of pastes made at a W/(C + SF) of 0.25 with 10 and 20% silica fume content has been measured.[18] Values of the coefficient of permeability after 180 days of curing was $>1 \times 10^{-13}$ m/sec for the specimen with 20% silica fume as well as for a blast-furnace slag cement containing 65% slag. The value for a sulfate-resisting paste was found to be 2×10^{-13} m/sec.

Relative diffusion rates of water vapor have been computed from data obtained by drying specimens at 11% relative humidity.[12] It has been found that a specimen made at a W/(C + SF) of 0.6 with 20% silica fume has a diffusion coefficient about 20% of that without the addition.

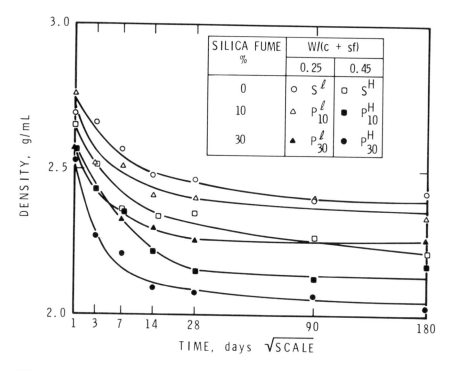

FIGURE 16.   Density of blends vs. time of curing using adjusted porosity values by water technique.

It is important to know the Cl⁻ diffusion rates through concretes exposed to deicing salts. Chloride diffusion data have been obtained for silica fume-cement pastes made at a W/(C + SF) of 0.15 to 0.225 with a silica fume content of 16%;[19] the diffusion coefficient varied between 0.7 and 1.9 × 10⁻⁹ cm²/sec. However, the value for blast furnace slag cement prepared at a W/C of 0.5 is less than this value[20] and that for sulfate-resisting cement paste formed at a W/C of 0.30 is between 19 and 25 × 10⁻⁹ cm²/sec.[18] The diffusion coefficient of the silica fume-cement paste is not as low as might be expected since its electrical resistivity is two orders of magnitude greater than that of normal cement paste.[18] It has been suggested that chloride ion diffusion is not a simple function of the porosity and that some "solid state" diffusion might occur.[18] In addition, in very low W/(C + SF) specimens a significant amount of interfaces between unhydrated and hydrated phases is present. Some interfaces involving Ca(OH)₂ may also occur in contrast to that in blast furnace slag. Chloride ion may migrate through these interfaces.

V. MECHANICAL PROPERTIES

A. Compressive Strength

Development of compressive strength with time by measurement of 50 × 50 mm cubes for six mixes, referred to in Table 1, is shown in Figure 17.[7] At a W/(C + SF) of 0.25 the strength for P_{10}^l is greatest (of the three mixes) between 28 and 90 days of curing, while S^l and P_{30}^l are similar up to 14 days. By 90 days, S^l has the highest strength, i.e., 124 MPa after 180 days of curing; P_{30}^l has the lowest, at 106 MPa.[7,18] The rate of strength development after 28 days is much greater for S^l than for the other two mixes. At a W/(C + SF) ratio of 0.45, P_{30}^H has the highest strength between 1 and 180 days;

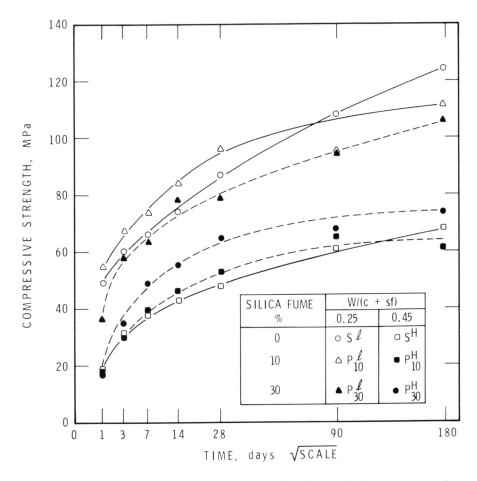

FIGURE 17. Compressive strength of cement pastes with different silica fume contents vs. hydration time.

however, the rate of strength development is greatest for S^H after 28 days, exceeding the strength of P^H_{10} between 90 and 180 days.

The greater rate of strength development for S^l and S^H between 28 and 180 days can be explained by the relative decrease in the hydration rate of the blends in this period. This is probably a result of the large decrease in permeability in the blends due to the reduction in $Ca(OH)_2$ content.[5]

B. Microhardness

The microhardness of a material is the value obtained by pressing a diamond or similarly shaped implement, the indenter, into the surface with a known load. The value, in load per unit area, has been found to be related to properties such as flexural and compressive strength of cement paste. Since indentations are very small, this technique serves as a useful tool for studying mechanical behavior nondestructively.

Porous materials tend to have a heterogeneous microstructure involving a wide range of particles and pore sizes and may also vary widely in composition. Microhardness measurements may thus vary from region to region and, to be representative of the bulk, many readings should be taken.

Development of microhardness with time for the six mixes is illustrated in Figure 18.[7] The spread of data for different silica fume contents at both values of W/(C +

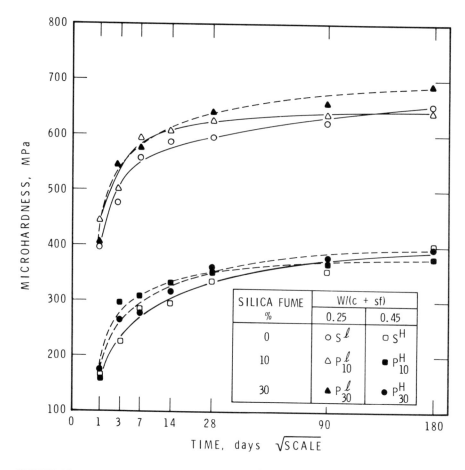

FIGURE 18.   Microhardness of cement pastes with different silica fume contents vs. hydration time.

SF) is much less than that observed for compressive strength. At a W/(C + SF) of 0.25, P^l_{30} develops the maximum hardness, i.e., 688 MPa after 180 days, while S^l and P^l_{10} have similar values of hardness of about 650 MPa after the same period. At a W/(C + SF) of 0.45, P^H_{10} shows the greatest microhardness up to 14 days. At 180 days, S^H and P^H_{30} have marginally higher values than P^H_{10} that shows the lowest microhardness. At both values of W/(C + SF) the specimens without silica fume show the greatest rate of increase in hardness between 28 and 180 days.

C. Young's Modulus

A method of measuring modulus of elasticity in flexure (stress per unit strain) by loading thin circular discs at their center with three equally spaced edge supports has been used with paste specimens.[21] Results for Young's modulus measured in this way as a function of curing time are shown in Figure 19.[7] At a W/(C + SF) of 0.25 the rate of increase of modulus is similar for the three mixes after 7 days of curing; values at 180 days are 2.92, 2.83, and 2.58 × 10⁴ MPa for P^l_{10}, S^l, and P^l_{30}, respectively; the compressive strength for P^l_{30} is also the lowest.

At a W/(C + SF) of 0.45 the rate of modulus development for S^H is maximum between 28 and 180 days, as for compressive strength and microhardness, and the value at 180 days is also the highest of the three mixes (the values being 1.75, 1.64, and 1.55

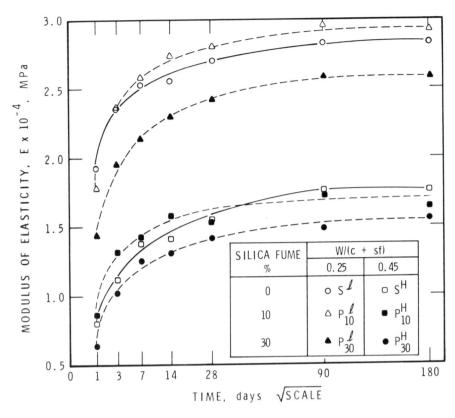

FIGURE 19.   Young's modulus of cement pastes with different silica fume contents vs. hydration time.

$\times$ 10⁴ MPa for S^H, P^H_{10}, and P^H_{30}, respectively). Microhardness and compressive strength values for P^H_{10} are, however, lowest. Modulus by resonance frequency determinations also indicate a similar trend showing increases with silica fume content up to 20% at a W/(C + SF) of 0.04 and 0.60, up to 60 days of curing.[12]

D. Factors Affecting the Development of Strengths

The porosity values determined by the water technique for different mixes are presented as a function of curing time in Figure 20. In all cases porosity decreases as curing time increases. At a W/(C + SF) of 0.25, the rate of porosity decrease is greatest for S^l and P^l_{10} prior to 28 days. After 28 days S^l shows the greatest decrease and the porosities for P^l_{30} and P^l_{10} become equal; at 180 days the values are approximately 18.9, 18.9, and 16.3 for P^l_{30}, P^l_{10}, and S^l, respectively.

At a W/(C + SF) of 0.45, the pastes without silica fume, S^H and P^H_{10}, show similar values for the first 14 days, but S^H shows the largest decrease in porosity after 28 days; values at 180 days are 28.8, 31, and 32.2 for S^H, P^H_{30}, and P^H_{10}, respectively. Prior to 28 days, P^H_{10}, S^H, and P^H_{30} have the lowest porosities in that order. These trends are similar to those for Young's modulus.

Porosity measurements (Figure 20) show that pastes without additions of silica fume have lower porosity values, because hydration reactions between 28 and 180 days continue at a faster rate in them. The difference in porosity, however, is not so large as would be expected from the difference in the extent of the reaction (assessed from the nonevaporable water content) because of the offsetting effect of density of products.

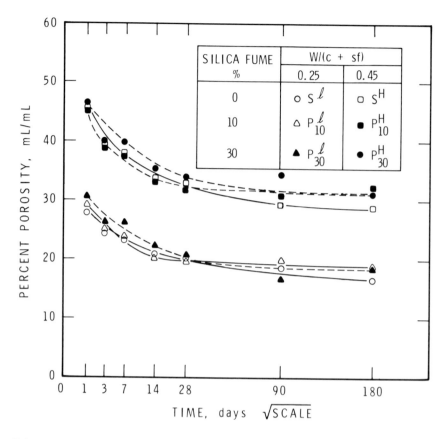

FIGURE 20.   Porosity determined by adjusted water technique for cement pastes with different silica fume contents vs. hydration time.

The density of C-S-H formed in the presence of silica fume is 2.05 g/mℓ, whereas its value when formed in the absence of silica fume is 2.24 g/mℓ. The C-S-H formed in the presence of silica fume would thus be more effective, per unit weight of nonevaporable water, in filling pores of the matrix.

   The value of the mechanical property of porous composite bodies, such as C/S fume blends, is dependent on several factors, including porosity, density of product, bonding between reaction products, and homogeneity. Although lower porosity could indicate a higher value for a mechanical property, other factors might have a countering influence. At a W/(C + SF) of 0.45, S^H does not have the highest compressive strength or microhardness after 180 days curing, despite lower porosity. The lower Ca(OH)$_2$ content of P^H_{10} and P^H_{30} (giving a more homogeneous body, possibly with fewer areas for crack initiation) can increase the strength of these paste blends. This factor is not so dominant with regard to Young's modulus, as may be seen in the results at a W/(C + SF) of 0.45 where the value for S^H is greater than either P^H_{10} or P^H_{30}; in this case the higher density product with the lower porosity appears to influence Young's modulus resulting in the higher value.

E. Porosity/Mechanical Property Relationships

   Several expressions have been used to describe the dependence of porosity on strength, fracture energy, and Young's modulus.[22-24] The following equation has often been used for describing mechanical properties:

Table 2
LINEAR REGRESSION ANALYSIS OF
COMPRESSIVE STRENGTH,
MICROHARDNESS, AND MODULUS OF
ELASTICITY VS. POROSITY RESULTS

Sample	S_o(MPa)	b_S	Correlation coefficient r
S^I	610.94	0.0960	0.98
P^I_{10}	371.5	0.0682	0.94
P^I_{30}	252.3	0.0539	0.92
S^H	448.7	0.0668	0.98
P^H_{10}	746.45	0.0804	0.96
P^H_{30}	669.88	0.0693	0.91
	H_o(MPa $\times$ 10^{-1})	b_H	
S^I	114.29	0.0329	0.91
P^I_{10}	117.49	0.0316	0.87
P^I_{30}	94.84	0.0199	0.90
S^H	137.09	0.0438	0.94
P^H_{10}	212.81	0.0555	0.87
P^H_{30}	182.39	0.0481	0.96
	E_o(MPa)	b_E	
S^I	4.35×10^4	0.0245	0.93
P^I_{10}	5.25×10^4	0.0311	0.95
P^I_{30}	4.16×10^4	0.0269	0.95
S^H	6.01×10^4	0.0417	0.93
P^H_{10}	4.90×10^4	0.0381	0.91
P^H_{30}	4.86×10^4	0.0363	0.88

$$M = M_o e^{-b} M^P$$

where M = mechanical property, M_o = mechanical property at zero porosity, P = porosity, and b_M is a pore shape factor.[22]

Plots of log mechanical property against porosity have been made for hydrated silica fume-Portland cement blend specimens cured from 3 to 180 days; the plots show six straight lines for each mechanical property.[7] The results for linear regression analysis, recorded in Table 2, show the intercept, the mechanical property at zero porosity (S_o, H_o, E_o, i.e., strength, microhardness, and Young's modulus, respectively), and the shape factor (b_S, b_H, and b_E) for each specimen. The correlation coefficient for each line is also given.

1. Compressive Strength

The values of S_o derived from compressive strength analysis at a W/(C + SF) of 0.25 decrease with increase in silica fume content. This system is very complex, especially since a large amount of unhydrated cement remains for S^I even at 180 days, and b_S values decrease with S_o values. At a W/(C + SF) of 0.45, S_o values for P^H_{10} and P^H_{30} are much higher than those for S^H, i.e., 746.5, 669.9, and 448.7 MPa, respectively. This is consistent with the fact that in P^H_{10} and P^H_{30}, Ca(OH)$_2$ content is greatly reduced, and despite the higher porosity for P^H_{10} and P^H_{30} at 180 days the strength for P^H_{30} was higher than that of S^H.

2. Microhardness

The values of H_o and b_H are also presented in Table 2. P_{10}^l has slightly larger H_o values than those of P_{30}^l and S^l at a W/(C + SF) of 0.25. The values of b_H change in the same manner as H_o. At a W/(C + SF) of 0.45, H_o varies as $P_{10}^H > P_{30}^H \gg S^H$, in the same order as for b_H, S_o, and b_S. This is consistent with the compressive strength results and explains why, despite the higher porosity of P_{10}^H and P_{30}^H in relation to S_H, the microhardness values after 180 days of curing do not differ from each other to any large extent.

3. Young's Modulus

The values for E_o presented in Table 2 show a different trend from those of compressive strength and microhardness. At a W/(C + SF) of 0.45, $S^H \gg P_{10}^H > P_{30}^H$. Modulus values for S^H are also greater than P_{10}^H and P_{30}^H at 180 days. At a W/(C + SF) of 0.25, no particular trend is evident, the values being $P_{10}^l > S^l > P_{30}^l$. The results appear to be dominated by the residual unhydrated cement.

The values of S_o, H_o, and E_o at a W/(C + SF) of 0.45 illustrate the response of the properties of the various composites to different mechanical stresses. For S_o and H_o, $P_{10}^H > P_{30}^H \gg S^H$, but for E_o, $S^H > P_{10}^H$ and P_{30}^H. Similar work has been reported for autoclaved C/S systems.[25] The values for the autoclaved 20% silica-cement mixture were similar to those for S^H reported here, but the values of the three properties, E_o, S_o, and H_o, of P_{10}^H and P_{30}^H were not different. The values of the ratio E_o/S_o for S^H, P_{10}^H and P_{30}^H are 133.9, 65.6, and 72.6, respectively, but the value of the ratio E_o/S_o varies from 110 to 186 for autoclaved C/S mixtures containing 10 to 50% silica. Silica fume addition results in the combination of a relatively low E_o, probably due to the low-density hydrate product, and a high S_o, due to low $Ca(OH)_2$ content and possibly to pore shapes leading to lower stress intensities.

F. Interrelationships Among Mechanical Properties

Linear plots of microhardness vs. compressive strength, Young's modulus vs. compressive strength, and Young's modulus vs. microhardness are presented in Figures 21 through 23, respectively, for a W/(C + SF) of 0.45.[7] Figure 21 shows good linear correlation between microhardness and compressive strength, and the slope of the line is independent of silica fume content. On the other hand, correlation between Young's modulus and compressive strength or microhardness, although good, shows different slopes depending on silica fume content. It is clear that the response of the Young's modulus measurement to changes brought on by silica fume is different from that of compressive strength and microhardness measurements. The reason is that the effect of $Ca(OH)_2$ crystals on the fracture process is different from the effect on stress-strain measurements.

G. Mechanical Property-Nonevaporable Water Relationship

Plots of the logarithm of the three mechanical properties vs. nonevaporable water content at a W/(C + SF) of 0.45 are presented in Figures 24 through 26. All are for specimens cured from 3 to 180 days. Correlation coefficients from linear regression analysis are all above 0.93, and for compressive strength and microhardness, the coefficients are above 0.95. The larger amount of nonevaporable water associated with specimens containing 0% silica fume for a similar gain of compressive strength or microhardness is evident from the comparison of slopes for those with 10 or 30% silica fume; that containing 10% silica fume has the greatest slope in the plot of strength and microhardness vs. nonevaporable water (Figures 24 and 25, respectively). Similar results may be observed for Young's modulus, except that the slopes for P_{10}^H and P_{30}^H are similar (Figure 26). At a W/(C + SF) of 0.25 there is relatively less difference between

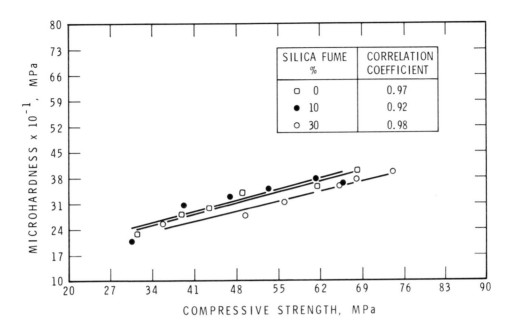

FIGURE 21. Microhardness vs. compressive strength for cement pastes containing silica fume (W/(C + SF) = 0.45).

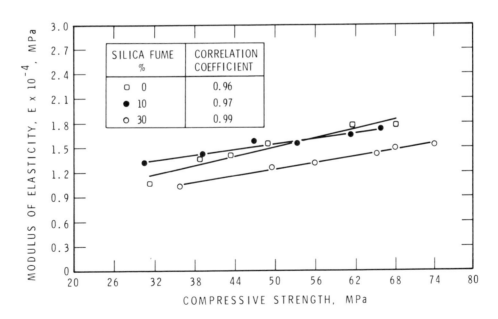

FIGURE 22. Young's modulus vs. compressive strength for cement pastes containing silica fume (W/(C + SF) = 0.45).

the slopes for S^i, on the one hand, and P^i_{10} and P^i_{30}, on the other, for the plot of the three mechanical properties. This is due to the increased influence of the unhydrated phase and the influence of the initial low W/(C + SF) on the degree of hydration.

The higher values of mechanical properties for P^H_{10} and P^H_{30}, with relatively low non-evaporable water content, as shown in Figures 24 through 26, are surprising. They are

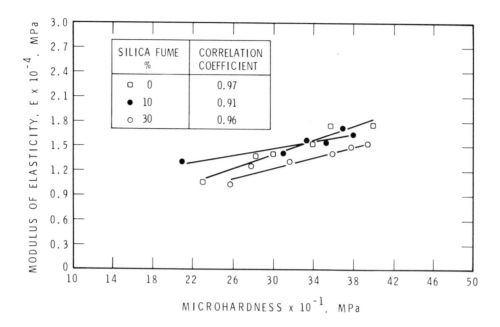

FIGURE 23.   Young's modulus vs. microhardness for cement pastes containing silica fume (W/(C + SF) = 0.45).

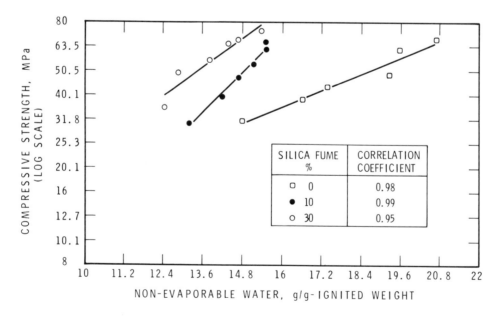

FIGURE 24.   Compressive strength vs. nonevaporable water content for cement pastes containing silica fume (W/(C + SF) = 0.45).

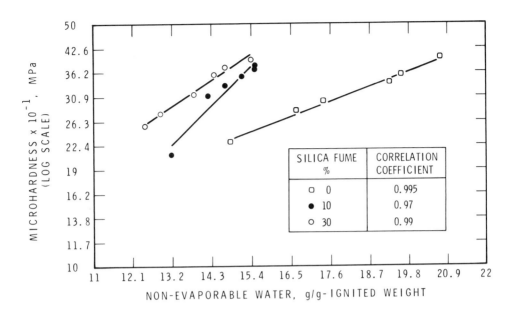

FIGURE 25. Microhardness vs. nonevaporable water content for cement pastes containing silica fume (W/(C + SF) = 0.45).

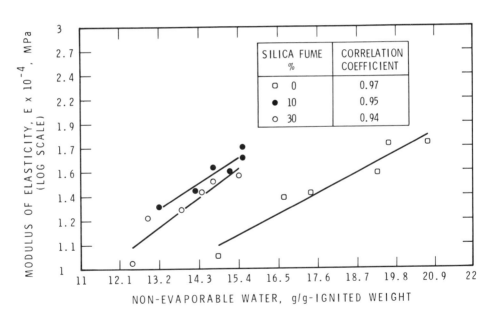

FIGURE 26. Young's modulus vs. nonevaporable water content for cement pastes containing silica fume (W/(C + SF) = 0.45).

probably due to the combination of a low-density, dispersed hydrate product of low CaO/SiO_2 ratio and to low $Ca(OH)_2$ content, leading to a relatively more homogeneous composite.

VI. DRYING SHRINKAGE

Drying shrinkage as a function of decreasing relative humidity is shown in Figure 27

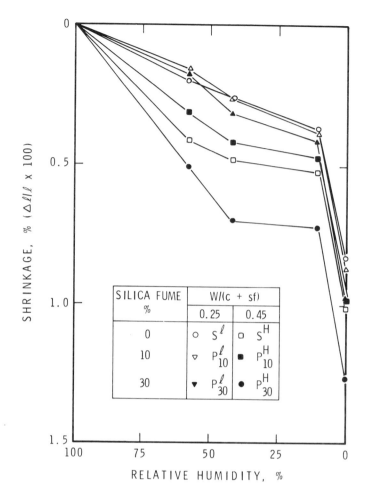

FIGURE 27.   Dependence of shrinkage of cement blend pastes cured for 90 days on relative humidity.

for various paste specimens.[6] For mixes formed at a W/(C + SF) of 0.25 no significant difference is observed for P_{10}^l and S^l, but P_{30}^l shows slightly larger values. Shrinkage for specimens S^H, P_{10}^H, and P_{30}^H is larger, therefore the silica fume content is greater, especially in the 100 → 42% relative humidity range; shrinkage in the 11 → 0% relative humidity range does follow this trend.

A large component of shrinkage in the 100 → 42% relative humidity range is usually observed to be irreversible and it appears as if 30% addition of silica fume at a W/(C + SF) of 0.45 significantly increases this form of shrinkage. The increase in drying shrinkage by the addition of 30% silica fume may be partially related to the fact that no calcium hydroxide is present to restrain the material during drying, and there is a greater amount of C-S-H per volume of material in P_{30}^H than in P_{10}^H or S^H. There is only a 10% difference in shrinkage during drying between 0 and 11% relative humidity for P_{10}^H and P_{30}^H, however, compared to over a 30% difference in the range 100 → 42% relative humidity. The difference appears to be due to the irreversible shrinkage effect, which may be caused by a difference in ionic charges on the C-S-H product; this may be caused by reduction of (Ca^{2+}) in the pore solution caused by complete depletion of $Ca(OH)_2$ in the material formed with 30% silica fume.

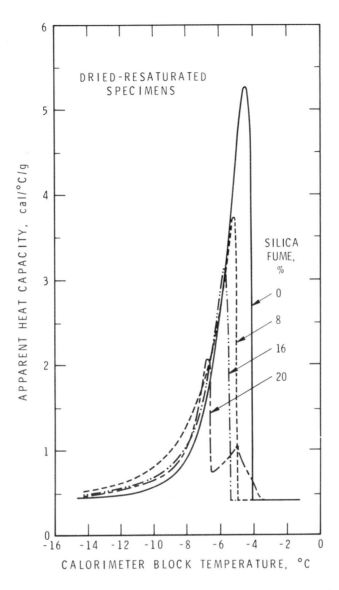

FIGURE 28.   Heat flow during cooling for virgin water-saturated pastes containing silica fume.

VII. DURABILITY

A. Freezing Action

Ice formation has been monitored at different temperatures with a low temperature adiabatic calorimeter in mature water-saturated samples prepared at a W/C of 0.6 with various silica fume contents.[12] Each paste was tested both in a virgin water-saturated state and in a resaturated state after having been dried for 3 days at 50°C at 33% relative humidity. The results for the undried pastes are shown in Figure 28. For the paste without silica fume, a large intensity peak representing ice formation takes place at about −4°C after a few degrees of super cooling.

Previous work has shown that water migrates out of small pores and freezes in locations of larger pores or outside of the specimen.[26] No such freezing was observed for

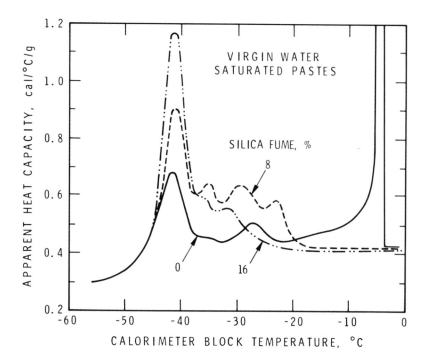

FIGURE 29.   Heat flow during cooling for dried, resaturated pastes containing silica fume.

the silica fume-containing pastes. Cooling below −20°C results in the formation of peaks for all the specimens, some of them due to freezing in smaller pores. The overall result is probably due to the discontinuous pore structure of the silica fume blend, causing low permeability of the paste which allows only slow migration of water. The results of the same experiment for the dried and resaturated specimens with different silica fume content are shown in Figure 29. These specimens show freezing peaks around −4 to −10°C. Drying at 50°C causes changes in the permeability of the specimens, probably due to accelerated aging. However, the pastes containing silica fume still have much lower intensity freezing peaks than those shown by the plain paste, reflecting the decreased rate of migration of water.

B. Frost Attack

The effect of frost action on pastes made with 0 and 10% silica fume at a W/(C + SF) of 0.60 without entrained air has been investigated[27] by exposing 127 × 25.4 × 25.4-mm prisms to freezing-thawing cycles consisting of freeze-in-air, thaw-in-water (−18 to +5°C), two cycles in 24 hr, according to ASTM standard test method C 666, Procedure B. The residual length change measured after thawing is presented in Figure 30 as a function of freezing and thawing cycles. It is shown that without silica fume expansion in bars exceeds 0.02% after five cycles and with 10% silica fume, after 18 cycles. After further freezing-thawing cycles both specimens deteriorate very rapidly. This indicates that 10% silica fume addition does not make a significant difference in the frost resistance of the cement paste. This is contrary to the results for mortars (Chapter 6) where it is found that silica fume addition improves the frost resistance for S/C ratios ⩾2.25.

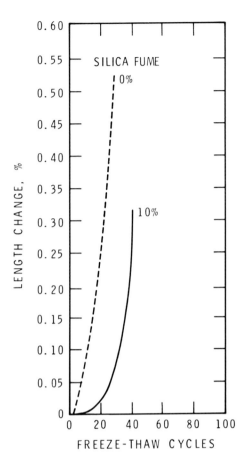

FIGURE 30.   Dependence of length change on freeze-thaw cycles of cement paste with and without 10% silica fume (14 days hydration W/(C + SF) = 0.60).

REFERENCES

1. Grutzeck, M. W. and Roy, D. M., Mechanism of hydration of Portland cement composite containing ferro silicon dust, Proc. 4th Int. Conf. on Cement Microscopy, 1982, 193.
2. Scheetz, B. E., Grutzek, M. W., Strickler, D., and Roy, D. M., Effect of composition of additives upon microstructure of hydrated Portland cement composites, Proc. 3rd Int. Conf. on Cement Microscopy, 1981, 307.
3. Regourd, M., Mortureux, B., and Gautier, E., Hydraulic reactivity of various pozzolanas, Proc. 5th Int. Symp. on Concrete Technology, 1981, 1.
4. Meland, I., Influence of condensed silica fume and fly ash on the heat evolution in cement pastes, Proc. 1st Int. Conf. on the Use of Fly Ash, Silica Fume, Slag and Other Mineral By-Products in Concrete, 1983, 677.
5. Huang, C.-Y. and Feldman, R. F., Hydration reactions in Portland cement-silica fume blends, *Cement Concrete Res.*, 15(4), 585, 1985.
6. Feldman, R. F. and Huang, C.-Y., Properties of Portland cement-silica fume pastes. I. Porosity and surface properties, *Cement Concrete Res.*, 15(5), 765, 1985.
7. Feldman, R. F. and Huang, C.-Y., Properties of Portland cement-silica fume pastes. II. Mechanical properties, *Cement Concrete Res.*, 15(6), 943, 1985.

8. Regourd, M., Mortureux, B., and Hormain, H., Use of condensed silica fume as filler in blended cements, Proc. 1st Int. Conf. on the Use of Fly Ash, Silica Fume, Slag and Other Mineral By-Products in Concrete, 1983, 847.

9. Traetteberg, A., Silica fumes as a pozzolanic material, *Il Cemento*, 3, 369, 1978.

10. Regourd, M., Mortureux, B., Aitcin, P. C., and Pinsonneault, P., Microstructure of field concrete containing silica fume, Proc. 4th Int. Conf. on Cement Microscopy, 1982, 249.

11. Feldman, R. F., Density and porosity studies of hydrated Portland cement, *Cement Technol.*, 3(1), 5, 1972.

12. Sellevold, E. J., Bager, D. H., Klitgaard-Jensen, E., and Knudsen, J., Silica fume-cement pastes — hydration and pore structure, in *Condensed Silica Fume in Concrete*, BML 82.610, Norwegian Institute of Technology, Trondheim, 1982, 19.

13. Beaudoin, J. J., Porosity measurements of some hydrated cementitious systems by high pressure mercury-intrusion — microstructural limitations, *Cement Concrete Res.*, 9, 771, 1979.

14. Feldman, R. F., Significance of porosity measurements on blended cement performance, Proc. 1st Int. Conf. on the Use of Fly Ash, Silica Fume, Slag and Other Mineral By-Products in Concrete, Vol. 1, 1983, 415.

15. Taylor, H. F. W., *The Chemistry of Cements,* Vol. 2, Academic Press, London, 1964.

16. Feldman, R. F., Pore structure damage in blended cements caused by mercury intrustion, *J. Am. Ceram. Soc.*, 62(1), 30, 1984.

17. Mehta, P. K. and Gjorv, O. E., Properties of Portland cement concrete containing fly-ash and condensed silica fume, *Cement Concrete Res.*, 12(5), 587, 1982.

18. Hooton, R. D., Permeability and Pore Structure of Cement Pastes Containing Fly Ash, Slag and Silica Fume, in press.

19. Preece, C. M., Frolund, T., and Bager, D. H., Chloride ion diffusion in low porosity silica cement paste, in *Condensed Silica Fume in Concrete*, BML 82.610, Norwegian Institute of Technology, Trondheim, 1982, 51.

20. Bakker, R., On the Cause of Increased Resistance of Concrete Made From Blast Furnace Cement to the Alkali Silica Reaction and to Sulphate Corrosion, D.Sc. thesis, Institut fur Gesteinstruttenkunde, RWTH, Aachen, 1980.

21. Sereda, P. J., Feldman, R. F., and Swenson, E. G., Effect of sorbed water on some mechanical properties of hydrated Portland cement pastes and compacts, Symp. Structure of Portland Cement Paste and Concrete, SP Rep. 90, HRB 1966, 58.

22. Ryshkewitch, E., Compressive strength of porous sintered alumina and zirconia, *J. Am. Ceram. Soc.*, 36, 65, 1953.

23. Schiller, K. K., Strength of porous materials, *Cement Concrete Res.*, 1, 419, 1971.

24. Roy, D. M. and Gouda, G., Porosity-strength relation in cementitious materials with very high strengths, *J. Am. Ceram. Soc.*, 56, 549, 1973.

25. Beaudoin, J. J. and Feldman, R. F., A study of mechanical properties of autoclaved calcium silicate systems, *Cement Concrete Res.*, 5, 103, 1975.

26. Helmuth, R. A., Capillary Size Restrictions on Ice Formation in Hardened Portland Cement Pastes, Vol. 2, National Bureau of Standards Monogr. No. 43, NBS, Washington, D.C., 1960, 855.

27. Feldman, R. F., The effect of sand-cement ratio on the frost resistance of silica fume-cement mortars, in press.

Chapter 6

PORTLAND CEMENT-SILICA FUME MORTARS

I. INTRODUCTION

Siliceous by-products such as condensed silica fume, fly ash, and blast furnace slag in combination with Portland cement produce on hydration a pore structure more discontinuous and impermeable than that of hydrated cement paste.[1-4] The difference in pore structure is related largely to the absence or low amounts of $Ca(OH)_2$ remaining intermingled with the C-S-H phase. This is discussed in an earlier chapter on the effect of silica fume on cement pastes.

In Portland cement mortars containing fine aggregates, such as silica sand, distinctly different properties are induced to the microstructure of the cement phase.[5] The degree of hydration is also reduced significantly.[6,7] Scanning electron microscopy (SEM) shows similar features which occur on or near the surface of a glass slide on the silica sand grain present in mortars.[8] These features include formation of a duplex film on the sand grain surfaces, development of large, well-formed $Ca(OH)_2$ crystals at intervals near the interface with their "C" axes roughly parallel to it, and development of stacked platelet secondary $Ca(OH)_2$ in open spaces immediately adjacent to the duplex film. These distinct differences in mortars produced by the presence of an interface largely involving the location of $Ca(OH)_2$ would lead one to anticipate significant differences in the properties of Portland cement mortars with and without silica fume.

This chapter will deal with the microstructure of mortar, focusing on the role of the interface, and will examine some of the physical properties that are affected by the interface. This chapter concludes with a discussion on the durability of mortars.

II. MICROSTRUCTURE BY SCANNING ELECTRON MICROSCOPY

The SEM examination of mortars can show the differences in microstructure of mortars containing various types of pozzolans. The microstructure of mortars containing slag shows some differences from those with silica fume.[4] In mortars containing 30% hydraulic slag (Figure 1A) and hydrated for 28 days, there is a separation between slag particles and the hydrated material made of tangled thin plates comprising Ca, Al, Si, and Mg. In mortars containing 25% slag and 5% silica fume (Figure 1B), the hydrated layer around slag is denser, poorly crystallized, and lower in Ca than that containing only slag. The decreased Ca content is related to the efficient reaction of silica fume with lime.

After 3 months of curing, the microstructural examination reveals that the slag grains are surrounded by a large hydrated layer. A good bond seems to have developed between the slag and C-S-H in the presence of silica fume (Figure 2). The microstructure of mortars containing nonreactive materials, such as crystalline slag or quartz, is also modified by silica fume. At 3 months of curing, compared to a porous structure containing a crystalline hydrated product (Figure 3A), silica fume promotes the formation of a denser looking structure containing amorphous C-S-H (Figure 3B).

Microstructural studies, in certain instances, can be used to explain the strength development in mortars containing various types of pozzolans. For example, in mortars containing fly ash, condensed silica fume, and volcanic ash, the chemical tests involving lime fixation did not correlate with strength tests.[4] In the presence of silica fume, C-S-H formed a compact mass with a vitreous-like appearance (Figure 4A). In

A

B

FIGURE 1.   (A) Mortar with 30% granulated slag; (B) mortar with
25% slag + 5% silica fume.

contrast to this, in the presence of volcanic ash (Figure 4B), the C-S-H formed a fibrous or a reticulated type structure. The compressive strength of mortar containing
30% silica fume was above 50 MPa, whereas that with 30% volcanic ash had a strength
of only 35 MPa.

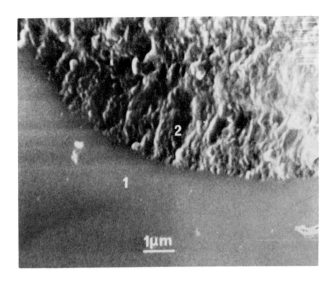

FIGURE 2. Mortar containing granulated slag + silica fume cured for 3 months (1) slag grain; (2) cement paste.

III. PORE STRUCTURE

A. Porosity and Pore Size Distribution

Very little work has been carried out on the pore size distribution of Portland cement-silica fume mortars. One such work[5] measured the change in pore distribution with a curing time of up to 180 days for mortars (sand/binder = 2.25) containing 0, 10, and 30% silica fume at W/(C + SF) of 0.45 and 0.60.

The results are shown in Figures 5 and 6 for W/(C + SF) of 0.45 and 0.60, respectively. Results for mortar without silica fume are shown in Figure 5A; the volume of pores of sizes down to 100 nm pore diameter for the 180 day specimen is about 5.5% of sample volume. This compares with less than 2.0% for the equivalent cement paste results of which were shown in Chapter 5. The volume of Hg intruded into the specimen decreases with curing time over the full range of pore sizes in a progressive manner, except for the 3-day specimen, which has a lower than expected porosity above 300 nm diameter. The curves for all samples at high pressures are of decreasing slope or concave to the pore size or pressure axis. (Pore diameter decreases as intrusion pressure increases.)

The pore-size distribution curves for specimens containing 10% silica fume are shown in Figure 5B. At values of diameter smaller than 60 nm, the total intruded volume decreases as the curing time increases. Specimens which have been cured for more than 3 days are of increasing slope or convex to the pressure axis at the high pressure end. Similar results have been observed in another study at 28 days.[9] This trend is similar to what has been observed previously for fly ash and slag blends,[1] where it was related to the disruption of discontinuous pores by the high pressure intrusion and to the extent of the reaction of the $Ca(OH_2)$.

At 100 nm the curves in Figure 5B for all but 1-day cured specimens merge with the same pore volume of about 7.5%, while from 200 to 3000 nm, the pore volumes for 90- and 180-day cured specimens are greater than all but the 1-day cured specimen.

The pore size distribution curves for specimens with 30% silica fume are presented in Figure 5C. The trend is the same as in Figure 5B in that in the 100 to 3000 nm range

A

B

FIGURE 3.   (A) Mortars containing quartz (B) or quartz and silica fume.

the total pore volume is greater in many cases for the specimens cured for longer times. The total pore volume down to a pore size of 300 nm for the 90- and 28-day cured specimens is greater than that for the 1-day specimen; the 3-day specimen has the lowest porosity. At the pore size of 30 nm, the total intruded volume decreases with age of curing and at ages greater than 3 days, the curves are of increasing slope or

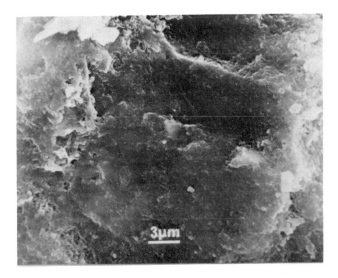

A

B

FIGURE 4. (A) Microstructure of mortars containing silica fume
and (B) volcanic ash.

sharply convex to the pressure axis. A plot of the slope of the curves in Figures 5 and
6 at the maximum intrusion pressure $10 \times dV/dlog\ D$ (%/nm) vs. age of curing is
presented in Figure 7. Designations for the various mixes are shown in Table 1. The
change in slope with time of these curves at maximum pressure (Figure 7) shows that
for B_{30}^i and B_{10}^i large changes occur up to 28 days, while for C^i the change is slight. This
is the period during which the major portion of $Ca(OH)_2$ is being formed by hydration
reactions or being consumed by the reaction with silica fume. Work with fly ash and
slag[1] has revealed that the convex distribution curves are a result of the formation of a
discontinuous structure when $Ca(OH)_2$ reacts with the pozzolan. It is concluded that

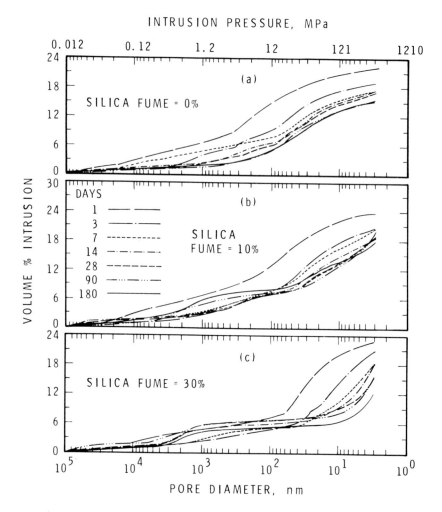

FIGURE 5. Pore size distribution curves of cement mortar with different silica fume contents (W/(C + SF) = 0.45). (a) 0% silica fume, (b) 10% silica fume, (c) 30% silica fume.[5]

during the Hg intrusion experiment, rupture of pore walls occurs at relatively high pressures. The entrances of these pores are of molecular dimensions but the bodies may be of the order of microns in size. Hg entering these ruptured pores at high pressures gives a false indication of a fine pore structure, whereas in fact the pores are relatively large but discontinuous. With silica fume, the pozzolanic reaction and the formation of the discontinuous structure commences after only about 3 days of curing.

The results for samples made at the W/(C + SF) of 0.60 (Figure 6) show a similar trend as for the W/(C + SF) of 0.45. The total volumes down to 100 nm pore diameter for specimens C^H, B^H_{10}, and B^H_{30} in Figures 6A, B, and C are approximately 3, 7, and 8%, respectively, for the 90-day cured specimens. Total porosity for the whole pore size range for C^H (Figure 6A) decreases generally with curing time in a progressive manner. Specimen B^H_{10} (Figure 6B) displays some merging in the distribution curves and at about 500 nm, samples cured for 14 and 28 days have total porosities that are lower than those for samples cured for 90 and 180 days. These effects are even more apparent for specimen B^H_{30}, where over some ranges 3-day cured specimens have lower porosities than

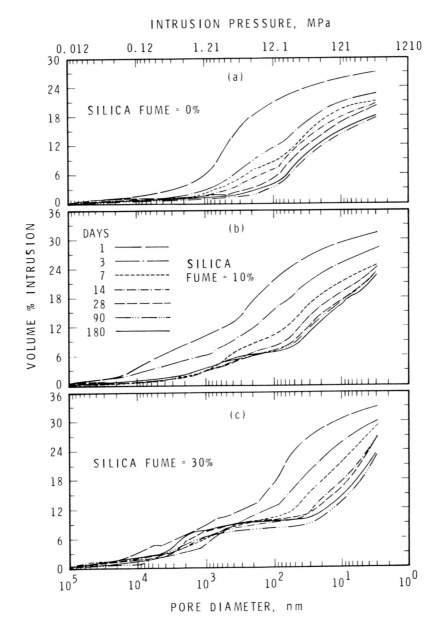

FIGURE 6. Pore size distribution curves of cement mortar with different silica fume contents (W/(C + SF) = 0.60). (a) 0% silica fume, (b) 10% silica fume, (c) 30% silica fume.[5]

those cured for longer periods; the 180-day cured specimen has as high a porosity as the 1-day specimen at 2000 nm. These effects are usually accompanied by abrupt jumps in the distribution curves. Both specimens B^H_{10} and B^H_{30} have an increasing slope at the maximum intrusion after about 3 days of curing. This effect, plotted in Figure 7, is similar to that for B^l_{10} or B^l_{30}.

The unusual form of pore structure development between 3 and 14 days, especially in the 100 to 1000 nm size range (Figure 6), may be the result of the $Ca(OH)_2$ concentrated around the sand grain forming a structure involving dissolution of some of the

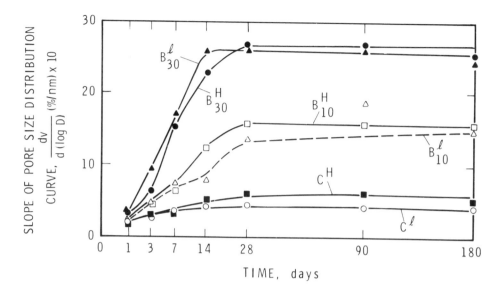

FIGURE 7. Dependence of slope of pore size distribution curve (at maximum intrusion pressure) on age and silica fume content.[5]

Table 1
DESIGNATION OF MIXES

| Silica fume (%) | W/(C + SF) | |
	0.45	0.60
0	● C^l	■ C^H
10	△ B^l_{10}	□ B^H_{10}
30	▲ B^l_{30}	● B^H_{30}

$Ca(OH)_2$ around the sand grain and then the formation of CSH by reaction between silica fume and $Ca(OH)_2$ that may dominate the pore size in the 100 to 10,000 nm range. The abrupt increases in intruded volume suggest that in this region of Hg pressure, rupture of the pore structure may occur. Such a rupture takes place at higher pressures for hydrated fly ash and slag blends, and this probably occurs with the silica fume blends at both intermediate and high pressures.[1,10] These abrupt increases in intruded volume at intermediate pressures have been observed in mortars and not in cement pastes containing silica fumes.[2]

The progressive manner (sequential reduction of porosity with time) in which the pore distribution changes with time for specimens C^l and C^H (Figures 5A and 6A) supports the idea that the lime that is formed preferentially around the sand grain is converted by reaction with nearby silica fume to C-S-H with a relatively discontinuous pore structure.

Changes in pore distribution curves in mortar blends containing silica fume are more marked than in paste blends (see Chapter 5).[5,11] This is especially true for the variations in the large pore-size range caused by the formation of discontinuous pores, due partially to the removal of lime accumulated at the sand-paste interface and partly to the reaction of line with silica fume within the matrix. Calcium hydroxide is formed in mortar in a variety of crystal sizes, but it is probable that larger crystals, or accumulations of crystals are formed around inclusions such as sand grains. In the absence of inclusions, as in paste blends, it is reasonable to assume that the occurrence in pastes of a large volume of trapped pores of about 3000 nm pore-size is unlikely.

The threshold values (the point where Hg intrudes at the maximum rate) of pore size for Hg intrusion in mortar blends and paste blends are quite different. In mortar blends the threshold value increases and becomes less abrupt with silica fume addition, while in paste blends it decreases.

Pore structure studies in mortars containing silica fume made at constant consistency with 0, 5, and 15% silica fume addition and cured for 15 days have also been carried out.[9,12] Increased pore volume in the 10^2 to 10^4 nm pore size range with increased silica fume content was observed.[12] Similarly increased pore volume in the 5×10^2 to 5×10^3 nm range was observed on field concretes made at W/C of 0.56 to 1.0 with 15% silica fume addition.[13]

B. Pore Discontinuity

The Hg reintrusion technique has been described in Chapter 5, Section III.D. This technique verifies to what extent initial Hg intrusion causes damage and this involves removing the Hg by distillation and repeating the Hg intrusion experiment. Hg reintrusions are performed on the six mortar mixes cured for 90 days and described in Table 1. Results are shown in Figures 8 and 9 (the curves for second intrusion are labeled with an R) and both the first and second intrusion curves are shown for each mix.[14] Results for specimens C^I and C^{II} (Figures 8A and 9A) show little difference between first and second intrusion, although the threshold pore diameters are slightly larger for the second intrusions. It may be observed from the first intrusion curve and from Table 2 that the pore volume, for specimens prepared at W/(C + SF) of 0.45, in the pore-size range 97,000 to 875 nm increases considerably with addition of silica fume: 1.65, 5.06, and 5.63% by volume for samples containing 0, 10, and 30% silica fume, respectively (Figures 8A to C). If the volume of pores in the 875 to 175 nm range is also taken into account, the pore volume in the range 97,000 to 175 nm is 3.90, 7.0, and 6.38% for the same samples. Hg reintrusion indicates that a significant portion of the pore volume of blends below 17.5 nm now falls in the 97,000 to 875 nm range. Volumes for the latter range are 2.86, 8.28, and 8.44% for additions of 0, 10, and 30% silica fume. Pore volume in the 97,000 to 175 nm range are 6.31, 9.48, and 9.40%, respectively.

Hg intrusion and reinstrusion results for specimens prepared at W/(C + SF) of 0.60 are shown in Figures 9A to C. The pore volumes in the 97,000 to 875 nm range are 0.94, 3.61, and 6.75% for additions of 0, 10, and 30% silica fume, respectively. Total volumes including those in the 875 to 175 nm range amount to 1.87, 6.37, and 8.22%. Reintrusion experiments confirm that a large portion of the pore volume formerly in the pore-size range <17.5 nm is shifted toward the 97,000 to 875 nm range. The value decreases from 11.58 to 6.51% for $B^{II}_{30}{}^{,90}$ in the smaller pore range. The sum of the pore volumes obtained by reintrusion for the 97,000 to 875 nm range and 875 to 175 nm range is 4.30, 7.34, and 12.15% for 0, 10, and 30% silica fume, respectively.

Comparison of these results with those of equivalent pastes shows that most of the changes between first and second intrusion occur for the mortar blend in the size range between 10^3 to 10^4 nm; for paste most changes occur gradually over the range 3.10^1 to 10^4 nm. Increase in volume in the higher size range is greater for mortar specimens.

IV. DRYING SHRINKAGE

The shrinkage of mortars containing silica fume has been measured by several workers and results are somewhat contradictory.[9,15,16] Results by Bager[16] are shown in Table

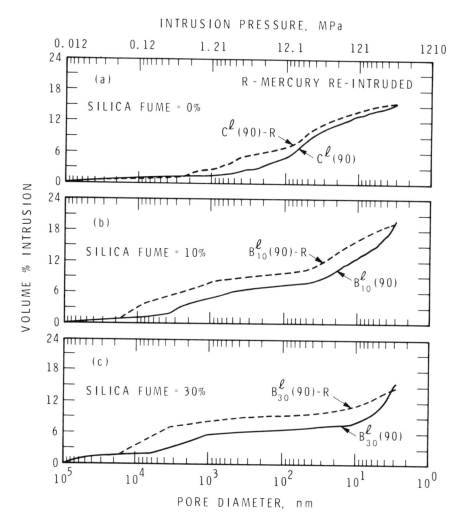

FIGURE 8.   Comparison of pore size distribution of cement mortar with different silica fume contents; first and second intrusions (W/(C + SF) = 0.45, 90 days curing).[14]

3 for different W/C ratios and silica fume contents (S). In this investigation W/C refers to W/C + 4S. Values for shrinkage after exposure to 40% relative humidity for 6 months are shown to increase with silica fume content at any W/C ratio and increases with W/C at 8 or 16% silica fume. No chemical admixtures were used in these mixes. Other researchers,[15] however, have measured shrinkage at a W/C ratio of about 0.4 and found that with 2.4% superplasticizer by weight of cement, shrinkage was substantially less in the presence of silica fume. Compare curve 1 (containing 40% by weight of cement of silica fume) and curve 2 (which contains no silica fume), shown in Figure 10. Curve 6 shows the shrinkage of the mortar containing no silica fume and no superplasticizer. As can be observed, there is a much larger increase in shrinkage with the addition of superplasticizer alone (sodium naphthalene sulfonate formaldehyde). Takagi et al.,[9] using W/(C + SF) of 0.55 and S/C (sand/cement ratio) of about 2, found a clear increase in drying shrinkage with silica fume content (Figure 11). In these mixes, however, superplasticizer was added to maintain a constant flow. At lower W/C (water/cement ratio), e.g., 0.25, silica fume additions do not seem to cause significant increases in drying shrinkage.

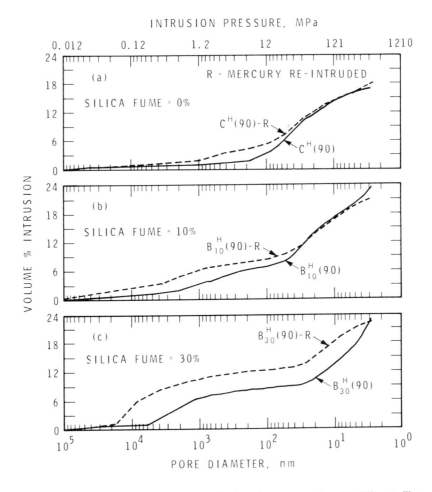

FIGURE 9. Comparison of pore size distribution of cement mortar with different silica fume contents, first and second intrusions (W/(C + SF) = 0.60, 90 days curing).[14]

V. PERMEABILITY

It has been all documented that even moderate additions of silica fume significantly improve the permeability of concrete,[17,18] and indications are that this is also valid for mortars,[19] although extensive work has not been carried out.

VI. MECHANICAL PROPERTIES

A. Compressive Strength

Compressive strength of mortars is improved by silica fume additions. Results by Takagi et al.[9] are shown in Figure 12 for mortars made at W/C ratios of 0.25 and 0.55 and S/C ratios of 2 and 1, respectively. Silica fume content varied between 0 and 40%. Improvements with the silica fume addition was as much as 15% after 28 days hydration. Other researchers[5,15] also observed improvements; for example, at a W/(C + SF) of 0.45, after 14 days curing the strengths were found to be 57, 44, and 40 MPa for 30, 10, and 0% silica fume, respectively. Corresponding strengths after 90 days curing were 77, 54, and 48 MPa. At a W/(C + SF) of 0.60, however, no significant increase in strength occurred with silica fume.

Table 2

PORE SIZE DISTRIBUTION OF MORTARS CURED FOR
90 DAYS BEFORE AND AFTER RE-INTRUSION (VOL %)

Pore diameter (nm)	C^I (90)	C^I (90)R	B^I_{10} (90)	B^I_{10} (90)R	B^I_{30} (90)	B^I_{30} (90)R
97,000—875	1.65	2.86	5.06	8.28	5.63	8.44
875—175	2.25	3.45	1.94	1.20	0.75	0.96
175—17.5	8.11	6.92	3.72	4.67	0.94	1.15
17.5—2.9	3.90	2.55	9.23	5.42	8.26	4.22
Total	15.91	15.78	19.95	19.57	15.58	14.77

Pore diameter (nm)	C^{II} (90)	C^{II} (90)R	B^{II}_{10} (90)	B^{II}_{10} (90)R	B^{II}_{30} (90)	B^{II}_{30} (90)R
97,000—875	0.94	1.91	3.61	6.48	6.75	10.95
875—175	0.93	2.39	2.76	0.86	1.47	1.20
175—17.5	10.29	8.12	8.28	6.48	3.02	3.76
17.5—2.9	5.61	4.30	8.70	6.70	11.48	6.51
Total	17.77	16.73	23.35	20.95	22.82	22.42

Table 3

DRYING SHRINKAGE (%)

W/C + 4S	SiO$_2$ Content (%)		
	0	8	16
0.4	0.14	0.13	0.18
0.5	0.11	0.15	0.20
0.65	0.09	0.12	0.23
0.8	0.08	0.16	0.24

A comparison of the ratio of compressive strengths of mortars to pastes containing 0 and 30% silica fume and cured at a W/(C + SF) of 0.45 for different periods are presented in Figure 13.[5] With no silica fume, the paste is stronger than mortar, having a strength of 68 MPa compared to 55 MPa for mortar at 180 days. The mixes containing 30% silica fume show the reverse trend. After 7 days curing, the curves representing mortar and paste begin to diverge; the mean strength of mortar after 180 days curing is 82 MPa compared to 74 MPa for paste.

Fracture studies reveal that fracture passes through the fine aggregate.[9] This is significant because in mortars without silica fume addition this does not normally occur. However, the strength of a composite material is not only dependent on the relative proportions of components (e.g., in mortar, the proportion of cement paste and sand), but also on the strength of the bond between the components. Although silica sand is denser and stronger than cement paste, the strength of mortar is normally lower than that of paste owing partly to a weak S/C bond. The addition of silica fume to the mortar appears to improve that bond.

The preferential deposition of Ca(OH)$_2$ in the interfacial zone[7,8,20] around aggregates (in mortars without silica fume) appears to weaken the composite. Therefore, the strength of mortar without silica fume decreases significantly in comparison to paste

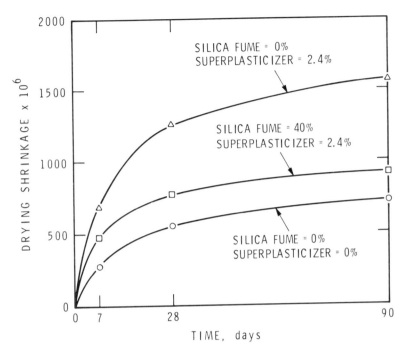

FIGURE 10. Drying shrinkage vs. time for mortars with and without superplasticizer and silica fume.[15]

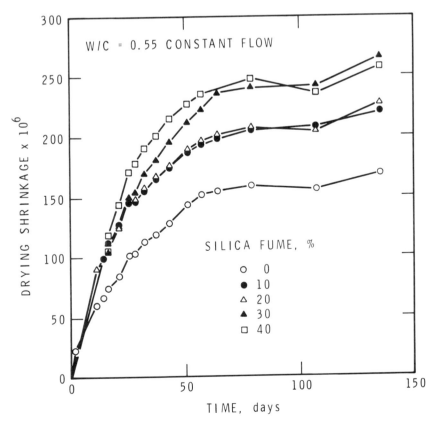

FIGURE 11. Drying shrinkage of mortars containing silica fume (W/(C + SF) = 0.55, constant flow).[19]

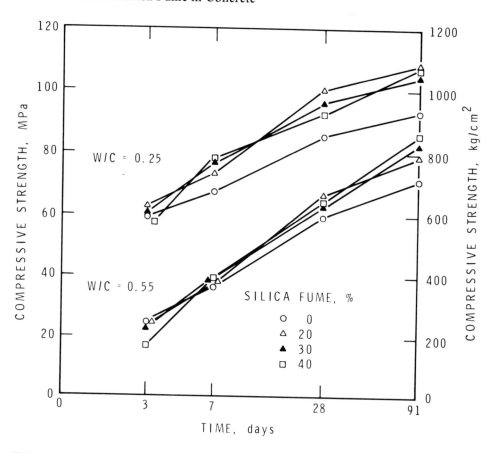

FIGURE 12.   Relationship between compressive strength and age at different W/(C + SF) and silica fume content.[9]

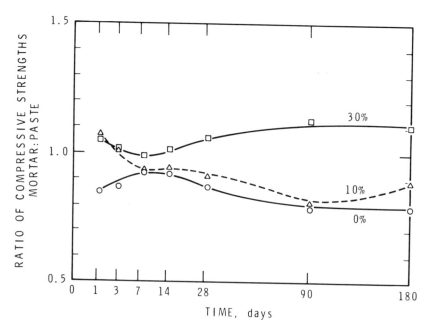

FIGURE 13.   Dependence of ratio of compressive strengths of mortar to paste on age and silica fume content (W/(C + SF) = 0.45).

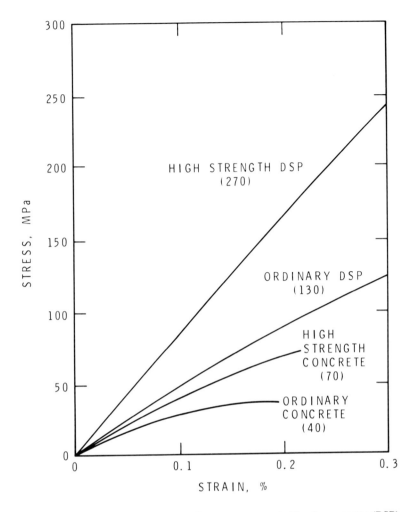

FIGURE 14.   Stress-strain plot for ordinary concrete and silica fume mortar (DSP). Compressive strengths in MPa are indicated by figures in brackets.[21]

after approximately 14 days of hydration when large quantities of Ca(OH)$_2$ are known to form (Figure 13). On the other hand, with 30% silica fume, after 7 days the strength of the mortar relative to the paste increases steadily. This is probably due to a better bond being formed between the sand grains by the new C-S-H formed from the reaction of Ca(OH)$_2$ with silica fume. The improvement of the aggregate-paste bond in this way facilitates fabrication of mortars with high strengths by the use of high strength aggregates and low W/C. A mortar strength of the order of 270 MPa has been achieved by using particles of calcined bauxite.[21]

B. Modulus of Elasticity

The modulus of elasticity of the low W/C mortar made with calcined bauxite aggregate (up to 4 mm) has been reported.[21] These results are shown in Figure 14. The value of modulus (108,000 MPa) calculated from these results for the dense mortar containing silica fume and bauxite, high strength DSP, and ordinary aggregates, ordinary DSP, clearly reflects the high density material as well as the effective bond between the paste and the aggregate.

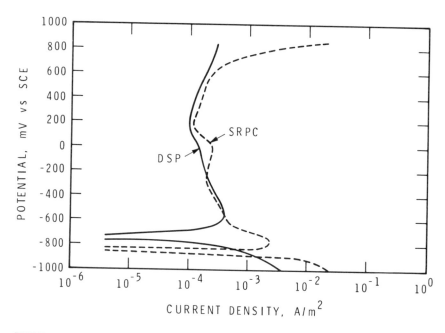

FIGURE 15.   Potentiostatic polarization curves for steel in dense silica fume blend mortar with W/(C + SF) = 0.15 and in sulfate resisting cement mortar with W/C = 0.50 in deaerated CO_2 saturated water at 22°C.[22]

VII. ELECTROCHEMICAL BEHAVIOR

The corrosion of steel embedded in mortars is an electrochemical process in which the cement pore water acts as the electrolyte. For corrosion to occur, the cement must allow (1) passage of electrical current, (2) diffusion of oxygen and water to the steel surface, and (3) transportation of the corrosion products away from the steel. The overall property of the mortar which controls its protective capabilities is its electrical resistivity. This, in turn, is modified by factors such as porosity and the chemistry of the pore water, particularly pH and chloride content. It is a combination of these properties which determines the corrosion rates.

Preece et al.[22] measured the potentiostatic polarization curves for a silica fume-Portland cement-water system prepared at W/(C + SF) of 0.15 to 0.20 (DSP) and compared them with curves obtained with a plain mortar made at W/C of 0.5 with sulfate-resisting Portland cement (SRPC). In deaerated $Ca(OH)_2$ solution, little difference was found between these mortars (Figure 15) in terms of the current passing, although the electrical resistivity of the silica fume containing mortar is known to be almost three orders of magnitude greater; it was concluded that when steel is not actively corroding the electrical resistivity is not a limiting factor.

In NaCl solution, there was a marked difference in the potentiostatic response of the same two mortars (Figure 16). The steel in the plain mortar was completely depassivated and corroded at rates up to three orders of magnitude greater than did the steel in the silica fume containing mortar. However, the onset of corrosion is probably exaggerated due to accelerated migration of chloride caused by the impressed potential.

The high electrical resistivity is considered a major factor for corrosion in practical situations; it has been concluded that this high electrical resistivity limits the corrosion due to the formation of galvanic cells at flaws in the mortar such as cracks or voids. Important conclusions could have been made if in this work[22] mortars prepared at the

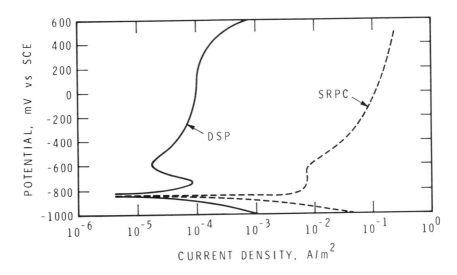

FIGURE 16. Potentiostatic polarization curves for steel in dense silica fume blend mortar (DSP) with W/(C + SF) = 0.15 and in sulfate resisting mortar (SRPC) with W/C = 0.50 in deaerated in NaCl solution saturated with Ca(OH)$_2$ at 22°C.[22]

same W/C as the silica fume had been subjected to potentiostatic polarization measurements.

VIII. DURABILITY

A. Frost Resistance

Recent work on mortars[12,14,23,24] and on concrete[25-27a] has shown that frost resistance is improved by incorporation of silica fume into the mortar mix. Other work on concrete has shown reduced frost resistance (see Chapter 9). Mortar mixes with a S/C ratio of 3:1 were made at a W/C of 0.48, 0.55, 0.60, and 0.70 by Traetteberg;[12] these mixes were made with 0, 5, 15, and 25% silica fume replacement of cement with the W/C ratio adjusted as W/(C + SF). Air entrainment and a combination of air entraining and plasticizing agent was added to some mixes. The results for change in dynamic modulus of elasticity determined by resonance frequency measurements due to freezing and thawing cycles (ASTM standard test method C 666 Procedure B; two cycles in 24 hr) are shown in Figure 17 for only up to 60 cycles. The plain mortar, prepared at W/C of 0.48, shows no deterioration but specimens made at W/C of 0.55 and over show a rapid decrease of modulus. Air entrainment is beneficial but the presence of air entrainment and superplasticizer appears to reduce the durability to some extent.

The addition of 5% silica fume improves considerably the durability of all the mortars except those made at a W/(C + SF) of 0.70. In other samples, little change in modulus is observed up to 30 cycles. With 15 and 25% silica fume additions the durability of mortar at W/(C + SF) of 0.70 also shows little change up to 30 cycles. These data were only for up to 60 freezing-thawing cycles. However, compressive strength of 4 mortar specimens were determined after 240 freezing-thawing cycles. These were made at W/(C + SF) of 0.55 without air entrainment; for specimens containing 0, 5, 15, and 25% silica fume strengths of 27.7, 44.9, 54.1, and 57.8 MPa, respectively, were reported.

The frost resistance of mortars with silica fume contents of 0, 10, and 30, S/C ratio of 2.25 and W/(C + SF) of 0.45 and 0.60 has been investigated by measuring the

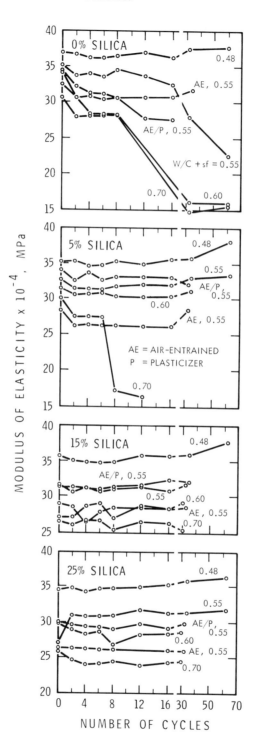

FIGURE 17.   Change of modulus of elasticity of
mortars with and without entrained air containing
0, 5, 15, and 25% silica fume.[12]

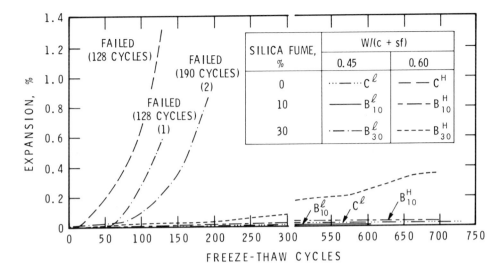

FIGURE 18. Expansion of mortars with different silica fume contents as a result of freeze-thaw cycle exposure.[14]

Table 4
SUMMARY OF RESIDUAL EXPANSION RESULTS OF
FREEZING AND THAWING TEST

Specimen no.	Cycles producing		Ultimate expansion (%)	Ultimate cycle	Expansion/ 100 cycles (% × 10²)
	0.1	0.2			
	% residual expansion				
1	2	3	4	5	6
C^l	≥732	≥732	0.012	733	0.16
B^l_{10}	≥722	≥722	0.007	722	0.10
B^l_{30}, 1	79	92	0.658	123	53.6
B^l_{30}, 2	102	124	0.848	190	44.6
C^H	36	54	1.168	128	91.2
B^H_{10}	≥696	≥696	0.028	696	0.40
B^H_{30}	350	560	0.346	696	5.0

residual expansion as a function of freezing-thawing cycles using similar freezing-thaw-ing cycling procedure as reported in Reference 12, but the specimens were exposed for more than 700 cycles[14] (see Chapter 5, Section V.I.2). Designations for the various mixes are included in the legend (Figure 18). Specimens C^l and B^l_{10} showed very little expansion even after exposure to in excess of 700 cycles. Two different preparations were tested for specimens B^l_{30}; both failed as shown in columns 2 to 6 (Table 4). This was not expected because specimen B^l_{30} displayed the greatest compressive strength and the lowest permeability of various specimens.[2]

Mixes prepared at W/(C + SF) of 0.60 displayed a somewhat different trend; the plain mortar specimen C^H expanded after only a few cycles. Neither specimen B^H_{10} or B^H_{30} failed even after 700 cycles although B^H_{30} showed higher expansion. Column 6 in Table 4 gives the average expansion × 10²/100 cycles. B^H_{10} with a value of 0.40 ranks

Table 5
PORE VOLUMES OF MORTARS IN PORE SIZE RANGES BENEFICIAL FOR FROST
RESISTANCE POROSITY, VOL %

Pore diameter (nm)	$C^I(90)$[a] (%)	$C^I(90)R$[b] (%)	$B^I_{10}(90)$ (%)	$B^I_{10}(90)R$ (%)	$B^I_{30}(90)$ (%)	$B^I_{30}(90)R$ (%)	$C^H(90)$ (%)	$C^H(90)R$ (%)	$B^H_{10}(90)$ (%)
20,000—2,000	0.63	1.30	3.23	5.47	2.49	5.75	0.40	0.75	1.47
2,000—350	1.22	4.11	2.61	2.71	2.06	1.48	0.52	1.70	3.40

Pore diameter (nm)	$B^H_{10}(90)R$ (%)	$B^H_{30}(90)$ (%)	$B^H_{30}(90)R$ (%)	$C^I(28)$[a] (%)	$B^I_{10}(28)$ (%)	$B^I_{30}(28)$ (%)	$C^H(28)$ (%)	$B^H_{10}(28)$ (%)	$B^H_{30}(28)$ (%)
20,000—2,000	2.85	4.00	8.45	0.25	0.96	2.20	0.30	0.78	4.58
2,000—350	2.75	2.09	2.07	2.40	2.35	3.10	1.08	3.65	3.42

[a]   (90) and (28) day curing.
[b]   R: after reintrusion.

third after B^I_{10} and C^I, while B^H_{30}, $B^I_{30.1}$, and C^H have values of 5.0, 53.6, and 91.2, respectively. It is clear that the addition of 10% silica fume at a W/(C + SF) of 0.60 without air entrainment provides considerable improvement in frost resistance of mortar exposed to freezing-thawing according to the particular procedure used.

However, it must be stressed that the results for these mortars were obtained using the less severe ASTM Method C666 Procedure B. In the equipment used the rate of cooling and warming conformed to C666 specifications but only 2 cycles/day were used. In addition in the preparation of mortars, Ottawa silica sand passing ASTM C109 was used. This is considerably finer than the sand normally used in concretes and will therefore decrease the average particle to particle separation for sand particles in concrete containing a comparable S/C ratio. The effect of S/C ratio is discussed in Section VIII.A.2.

1. Mechanism of Improved Frost Resistance Due to Silica Fume

The conventional method of rendering concrete frost resistant is to entrain air bubbles of the 10,000 to 100,000 nm diameter size in the concrete. These bubbles should be spaced no more than 200,000 nm apart. Usually in the order of 5 to 7% entrained air by volume of the concrete is found to be satisfactory. Litvan[23] has suggested that pore volume contained by pores in the size range 2000 to 350 nm as well as in pores >10,000 nm are effective in producing frost resistant mortar. This may explain the greatly improved frost resistance of specimens B^H_{10} and B^H_{30} over that of C^H discussed here. Pore volumes in the 20,000 to 2000 and 2000 to 350 nm pore diameter range are shown in Table 5 for both first and second Hg intrusion. Pore volumes residing in these ranges are greatly increased with the addition of silica fume as shown after first intrusion; after second intrusion the values are further increased. There is evidence that $Ca(OH)_2$ is preferentially deposited in the interfacial zone around inclusions such as aggregates in mortar and concrete.[8,20] Silica fume in these mixes reacts with $Ca(OH)_2$, thereby affecting pore distribution in the mortar and creates further pores in the range 20,000 to 350 nm at the interfacial zone around sand grains. Assuming that pores of the above size range form around homogeneous sand grains of uniform size (0.5 nm) in a mortar having a 2.25:1 sand-to-binder ratio, a simple calculation can demonstrate that spacing between these pores is less than 0.1 mm. This conforms to specification ASTM C457-71 recommended for frost durability. In addition, part of the pore volume in the above ranges is of the ink bottle type and relatively inaccessible to water. This property may be important in maintaining a low level of saturation in these pores during the freezing-thawing cycles, allowing them to act as reservoirs for water migrating from small pores during the freezing process.

Specimen C^I containing no silica fume also performed well. In this mix the pore volume in the range 20,000 to 350 nm increased from 1.85 to 5.41% by volume (Table 5) when Hg was reintruded, indicating that a large portion of these pores is relatively inaccessible. The mixture C^H which had corresponding values of 0.92 and 2.45% for Hg intrusion performed poorly. In addition to this factor, the lower W/C of C^I, 0.45, would ensure a lower general permeability and less capillary water than those for specimen C^H.

The rapid deterioration of mortar specimen B_{30}^I with freezing-thawing cycles (although possessing the highest strength and salt solution resistance[2,5]) appears to be anomalous on first examination. It is probably related to its low permeability and large silica fume content (30%). At 30% the silica fume content is in excess of that needed for complete reaction, resulting in a higher effective water:binder ratio. In addition, hydration is not completed, and at a W/(C + SF) ratio of 0.45 a relatively large amount of evaporable water still remains in the pores.[5] It is entrapped during the freezing-thawing cycles as a result of the low permeability of the specimen and promotes damage. The greater permeability of specimens B_{30}^H and B_{10}^H ensures greater frost resistance. It has also been found that relatively larger amounts of silica fume in concrete result in decreasing frost resistance (see Reference 5, Chapter 9).

2. The Effect of S/C Ratio and Silica Fume on the Microstructure and Freezing-Thawing Durability

The description of the mechanism for frost resistance of mortar containing silica fume in the previous section emphasizes the role of the silica fume-fine aggregate interfaces. It is postulated that the occurrence of space at or near the interface will act as a relief valve as in the case of entrained air, the mean distance between the spaces being similar in the spacing factor. In mortar, the greater the S/C ratio, the less the mean interface. Separation and this factor should have an influence on durability.

Measurements of pore size distribution by Hg intrusion have been made on mortars cured for 28 days with and without 10% silica fume addition.[28] These results are shown in Figures 19 and 20, respectively. To compare the structure of the paste phase for specimens containing different quantities of sand (ASTM C 109), the pore volume was computed per (volume of mortar) − (volume of sand). This parameter would include the volume of pores, if any, at the interface of the cement paste and sand. The volume of pores in the pore-diameter range of 20,000 to 2000 nm, for the specimens containing 10% silica fume increases with S/C ratio up to a value of 5.0% with the S/C ratio 3.0. With 0% sand, the volume is less than 1%. There is a large increase in the volume of pores between 2000 and 800 nm, and this increase in volume of pores also increases with the S/C ratio; the total volumes of pores down to 800 nm diameter are 2.5, 3.4, 4.0, 4.8, 4.1, 4.5, 6.0, and 11.6% for S/C ratios of 0, 0.5, 1.0, 1.5, 1.8, 2.0, 2.25, and 3.0, respectively. The total volumes of pores down to 800 nm diameter for mortars, on the other hand, without the silica fume addition are 0.8, 0.8, 3.3, 2.4, 3.2, and 5.3% for S/C ratios of 0, 1.5, 1.8, 2.0, 2.25, and 3.0, respectively; there is thus a large increase in pore volume down to 800 nm with the silica fume addition for most S/C ratios, and especially for those with ratios 2.25 and 3.0.

Residual length changes as a function of the total number of freezing-thawing cycles experienced by the above specimens are shown in Figure 21. All mortars, without silica fume up to a S/C ratio of 2.25 exceed 0.02% after relatively few cycles (8 cycles). The addition of silica fume extends this to just over 20 cycles for all S/C ratios except those

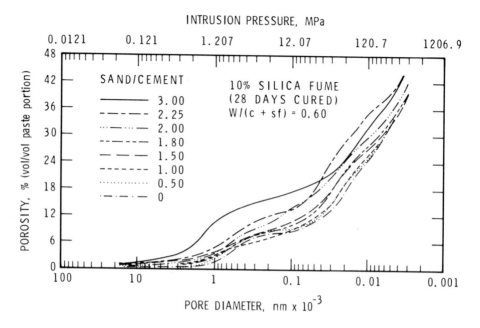

FIGURE 19.   Pore size distribution of mortar as a function of S/C, prepared at W/(C + SF) of 0.60 containing 10% silica fume. Intruded volume based on volume of paste phase.[28]

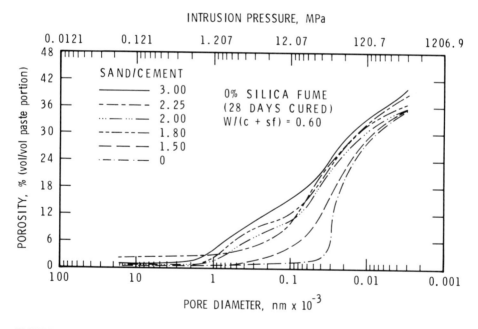

FIGURE 20.   Pore size distribution of mortars as a function of S/C prepared at W/(C + SF) of 0.60 without silica fume. Intruded volume based on volume of paste phase.[28]

of values 1.5 and 1.8, for which only about 15 cycles are required. However, when the S/C ratio is 2.25, 0.02% expansion occurs only after 400 cycles; at the S/C ratio of 3.0, expansion is only 0.011% even after 540 cycles. At an S/C ratio of 3, the mortar without silica fume expands by 0.02% after about 215 cycles, but the specimen had lost over 9% of its weight after 240 cycles. However, specimens with silica fume show no weight loss even after 540 cycles.

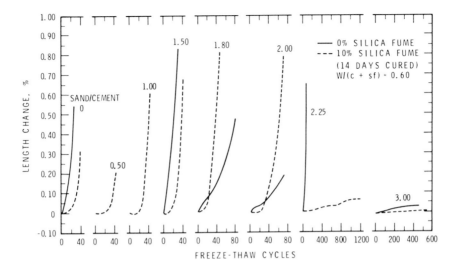

FIGURE 21. Expansion of mortars with different S/C and with 0 and 10% silica fume as a result of freeze-thaw exposure.[28]

These results indicate that silica fume addition does not significantly improve the frost resistance of pastes and of mortars unless the S/C ratio is 2.25 or over.

B. Resistance to Salt Attack

Several studies have been carried out on the effectiveness of silica fume (in mortars) to improve resistance to attack by aggressive agents.[29-31] Observations on the modifications of the pore structure with resultant effects on permeability and the chemical combination of a Ca(OH)$_2$ would mean that silica fume additions enhance chemical resistance of mortars.

1. Exposure to NH$_4$NO$_3$

Popovic et al.[29] studied the effect of 10% ammonium nitrate on three cements with and without 15% silica fume addition. The cements were ordinary Portland cement or that blended with 20% blast furnace slag, or that blended with 15% of natural pozzolan. The silica fume was of the high purity and activity type obtained from silicon metal production. Specimens were made at a constant fluidity and cured for 28 days. Compressive strength results for the ordinary Portland cement and blast furnace-Portland cement blend are presented in Figure 22. Curves are presented for mixes containing no additions, only superplasticizer, only silica fume, or silica fume and superplasticizer. Since these specimens were prepared at a constant consistency, the utilization of silica fume without superplasticizer would not be expected to perform as well as others, especially in the early periods of exposure. Results presented in Figure 23 show compressive strengths of specimens exposed to NH$_4$NO$_3$ and based on the strength of similar preparations stored in water. The samples with silica fume and superplasticizer are compared with those containing no admixture. It can be shown that silica fume addition greatly improves the resistance of both blended cements to the NH$_4$NO$_3$ solution; similar results were also obtained with the Portland cement-silica fume mixture.

X-Ray diffraction and Hg intrusion measurements showed that without silica fume, a substantial decrease in Ca(OH)$_2$ content occurs together with a large increase in porosity. However, when silica fume was present, the porosity did not change significantly on exposure to NH$_4$NO$_3$ solution.

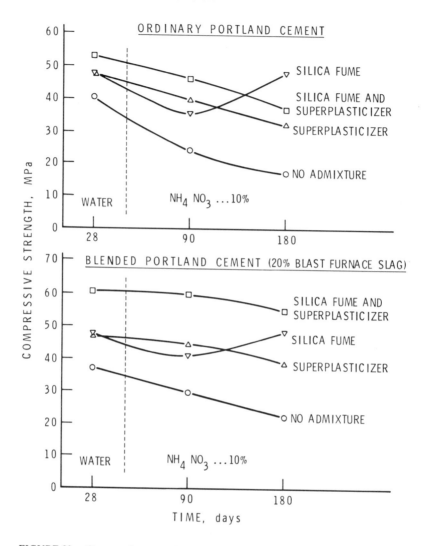

FIGURE 22.   Compressive strength of mortars with various additives as a function of time exposed to 10% NH₄NO₃ solution.[29]

2. Exposure to Sulfates

Specimens similar to those exposed to 10% NH_4NO_3 were exposed to 10% $(NH_4)_2SO_4$ solution.[29] Flexural strengths on exposure to these solutions is presented in Figure 24 for ordinary Portland and blast furnace Portland cements. In most cases, corrosion due to $(NH_4)_2SO_4$ is more severe than that due to $(NH_4)NO_3$. Silica fume addition improves durability. It was also observed that although porosity was not changed after the $(NH_4)_2SO_4$ attack, the strength loss was considerable. This may be the result of stresses and cracking created by the sulfate reaction with the aluminates and that a significant amount of ettringite fills many of the pores vacated by leaching of $Ca(OH)_2$ by the sulfate. Also, nitrate forms a complex with C_3A which is not as expansive as ettringite; the removal of lime by NH_4NO_3 solution also does not cause expansion.

Silica fume addition also improves the durability of the mortars to sulfate attack. This is confirmed by the results obtained by immersing mortars prepared at W/C of 0.60 in 10% $Na_2SO_4 \cdot 10H_2O$ solution; flexural strength changes, as a ratio of the

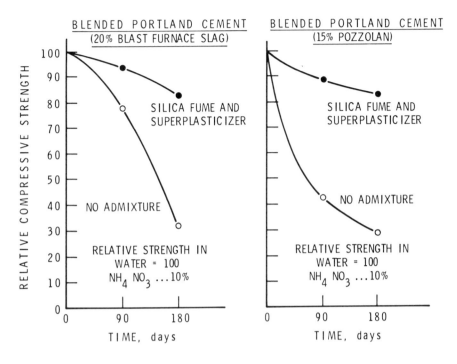

FIGURE 23. Ratio of compressive strength of corroded to uncorroded mortars with various additives in 10% NH₄NO₃ solution.[29]

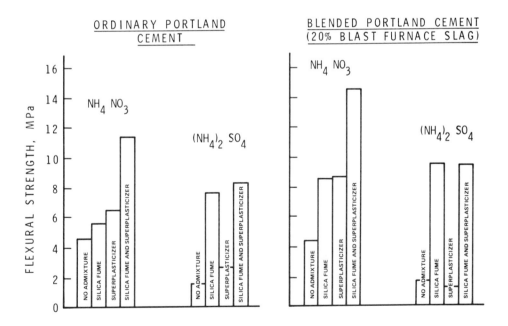

FIGURE 24. Flexural strengths of mortars with various additives after 90 days of exposure in 10% NH₄NO₃ or 10% (NH₄)₂ SO₄ solutions.[29]

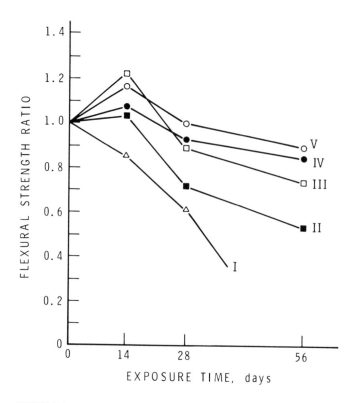

FIGURE 25.   Ratio of flexural strength of corroded to uncorroded mortars with various additives, as a function of exposure time in Na₂SO₄ solution. (I) Portland cement, (II) no. I + 45% slag, (III) sulfate resisting cement, (IV) I + 10% silica fume, (V) no. II + 10% silica fume.[29]

values with respect to those stored in water, are presented in Figure 25. Ordinary and blended Portland cement mortar with added silica fume exhibit better durability than sulfate-resisting cement (3.5% C₃A).

Durability was also measured as expansion of mortar bars, made at W/C of 0.485 and S/C ratio of 2.75 by immersing in 4% Na₂SO₄ solution.[31] These bars contained a 30% silica fume replacement and the cements contained 14.6, 13.1, and 9.4% C₃A. Expansion at 365 days was considered insignificant for the three cements. Of ten pozzolans tested in this work, the silica fume was ranked the best.

3. Exposure to Chloride Solutions

Mortars containing 10 and 30% silica fume replacement were exposed to a chloride solution containing 27.5% CaCl₂, 3.9% MgCl₂ and 1.2% NaCl. The mixes are those referred to in Table 1, and they were cured either for 7 or 28 days before exposure. Durability was monitored by measuring deflection in flexure of thin (6.4-mm thick and 75 mm in diameter) discs. Pore size distribution, Ca(OH)₂, and nonevaporable water contents were also determined. It was found that Ca(OH)₂ in the mortars without silica fume, as in the case of NH₄NO₃ solution exposure, was greatly reduced. This resulted in increases of porosity of specimens in the (97 to 2) × 10³ nm pore diameter range. A correlation between the increase in porosity in this range and the increase in flexural deflection is shown in Figure 26. It may be observed that the specimens with the lower W/(C + SF) (0.45), cured for longer periods, and containing 30% silica fume offered the greatest resistance to attack, i.e., they showed the lowest deflection. Pore size dis-

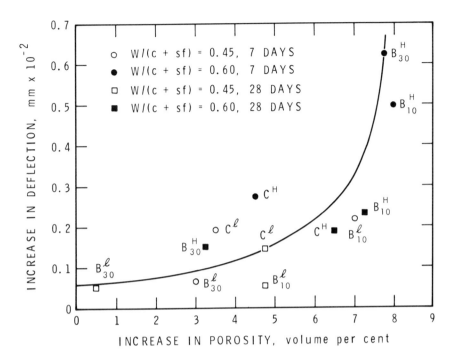

FIGURE 26. Dependence of increase in flexural deflection on porosity ((97 to 2) × 10³ nm) for cement mortars with various additives after exposure to salt solutions (≈360 days).[30]

tribution measurements on specimens prepared at W/(C + SF) of 0.6, cured for 28 days with 0, 10, and 30% silica fume, before and after exposure to salt solutions is presented in Figure 27. Large increases in porosity in the (97 to 2) × 10³ nm range are apparent, and this decreases with silica fume content. Another observation is that the total porosity which increases slightly with C^H [28] actually decreases with silica fume content and a simultaneous increase in evaporable water content occurred. This leads to the conclusion that reactions involving salt solution, silica, and low calcium C-S-H may be taking place; a product appears to be formed that has a pore-filling capability.

C. Effect of Silica Fume on Alkali-Aggregate Expansion

The most effective method of controlling expansive alkali-aggregate reaction is to reduce the total alkali content of the concrete below a critical limit. This method, however, is not always economical. Another method which has shown promise is to blend an approved mineral admixture with the high alkali cement.

Silica fume has been tested as one of the mineral admixtures for the purpose of reducing expansion due to alkali aggregate reaction. Oberholster and Westra,[32] using the Pyrex glass mortar prism test (ASTM C441-69), found that with silica fume as a blend, a "shrinkage" of 0.008% occurred. This was the lowest "expansion" of eight mineral admixtures used. In order to verify the effectiveness of the mineral admixtures for use of a natural reactive aggregate, a batch of quarry aggregate consisting of hornfels and graywackes from the Tygerberg Formation of the Malmesbury Group was selected. Each of the pozzolanic admixtures including silica fume was blended in quantities equal in volume to 5, 10, 15, 20, 25, and 30% by mass of cement (containing 0.97% equiv. Na_2O). The aggregate-cement ratio was 1.5. In order to minimize the dilution effect, this cement (0.97% equiv. Na_2O) was blended with a cement containing 0.16% Na_2O so that cements were obtained containing the same Na_2O equivalent as

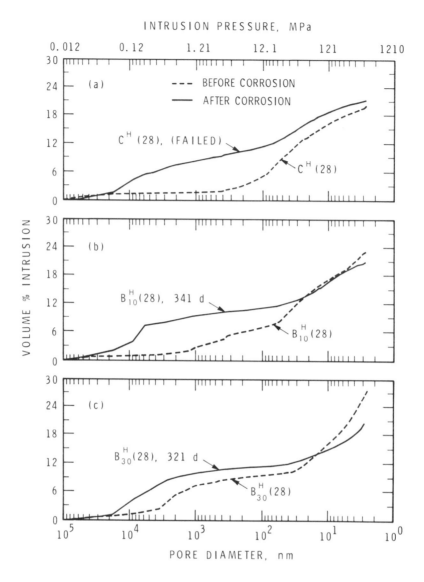

FIGURE 27.   Comparison of pore size distribution of cement mortar before and after corrosion (W/(C + SF) = 0.60, cured 28 days). (a) C^H, (b) B^H_{10}, (c) B^H_{30}.

the balance of the cement after replacement with the pozzolans. The expansion results of mixtures containing various admixtures after 555 days are shown in Figure 28. The top curve shows the effect of dilution with low alkali cement, which is a continual decrease in expansion as equivalent Na_2O decreases from 0.97 to 0.68 at 30% replacement. All the admixtures thus show a decrease in expansion considerably lower than this curve and their effect is not just the result of dilution. Assuming 0.1% expansion as the criterion for allowable expansion, a 10% by volume replacement of silica fume would be required, and silica fume was found to be the most effective of the mineral additions tested.

Hypotheses concerning the mechanism by which pozzolans provide resistance to alkali-aggregate reactions are generally centered on the type of C-S-H formed during hydration of cement. When the CaO/SiO_2 ratio of the C-S-H formed is approximately

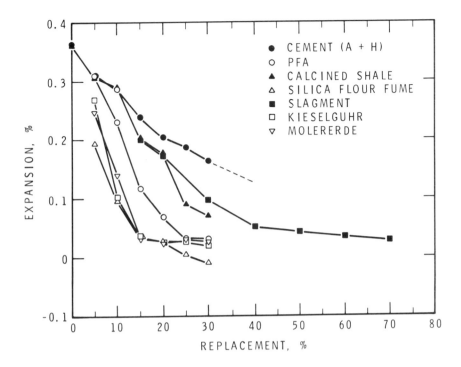

FIGURE 28.   Effect of replacing cement H (0.97% Na₂O) with increasing amounts of cement
A (0.16% Na₂O) or various mineral admixtures, on the linear expansion of mortar prisms
stored at 38°C for 555 days.[32]

1.2 or lower, this product it is suggested,[4] has an increased capacity for accommodating
Na₂O and K₂O in its structure, thereby reducing the hydroxyl ion concentration. During normal hydration without the mineral admixture this ratio is about 1.5.

Olafson,[33] applying a version of ASTM C227 modified by Brotschi and Mehta,[34] used crushed Pyrex as the aggregate and showed that when the pozzolan was of a high surface area type such as silica fume, lower quantities of addition were needed to contain the reaction. The results in Figure 29 show expansion as a function of the total CaO/SiO₂ for the three unhydrated cements with added silica fume. Thus, it can be seen that a CaO/SiO₂ of 1.9 to 2.3 is needed in order to keep expansions under 0.1%, depending on the type of cement. To attain this ratio 10 to 13% of silica fume should be added. (The more silica fume added, the lower the ratio will become.) The results of longer term expansion measurements using an Icelandic sand (Bjorgun) and a cement with 1.39% equiv. Na₂O alkali is shown in Figure 30. With both 7½ and 10% silica fume, the expansion after 3 years is about 0.06%.

Perry and Gillott[35] have also shown the effectiveness of silica fume to control alkali silica reactions by using Pyrex as the reactive aggregate suggested in ASTM methods C227 and C441. In addition, they tested a very reactive opal from Nevada according to ASTM C227 except that temperatures of 25° and 50°C were used in addition to the standard 38°C. The amounts of cement replaced by the silica fume ranged from 0 to 40% by weight.

Results of experiments performed at 50°C are shown in Figure 31. Replacement of cement by silica fume significantly reduced expansion, but it appeared that 20% replacement might be required so that the reaction of the opal could be controlled. Evidence presented in this work indicated, however, that superplasticizer addition at 15% replacement level of silica fume may influence expansion in a negative manner.

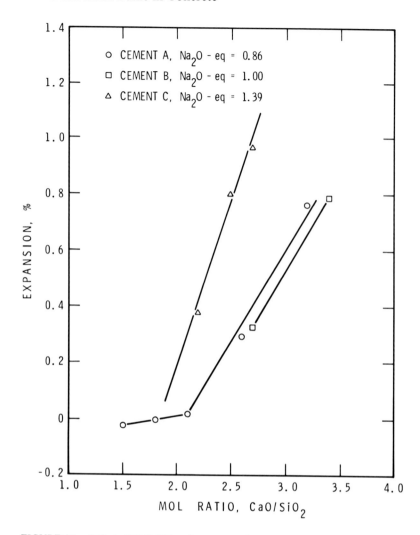

FIGURE 29.   Effect of CaO/SiO₂ of cement on linear expansion due to alkali-aggregate reaction.[33]

   Direct measurements of pore solutions using a high-pressure press technique has
demonstrated how as low as 5% silica fume reduced the hydroxyl ion concentration to
levels below 0.3 mol/ℓ.[36] This should be sufficient to contain most alkali-silica reactions, however, longer term studies are needed for confirmation. There is some evidence from results obtained after 200 days suggesting that the alkali-carbonate reaction
may not be controlled by as much as 25% silica fume addition.[37]

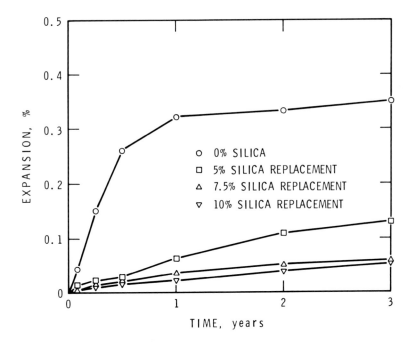

FIGURE 30.   Effect of silica fume on linear expansion of mortar bars containing Iceland sand and cement containing 1.39% Na$_2$O.[33]

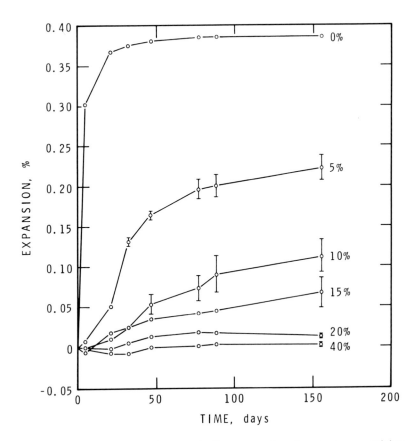

FIGURE 31.   Effect of silica fume on the linear expansion of mortar bars containing 4% reactive opal (specimens at 50°C and 100% relative humidity).

REFERENCES

1. Feldman, R. F., Significance of porosity measurements on blended cement performance, Proc. 1st Int. Conf. on the Use of Fly Ash, Silica Fume, Slag and Other Mineral Byproducts in Concrete, Montebello, Canada, 1983, 415.
2. Feldman, R. F. and Huang, C.-Y., Microstructural properties of blended cement mortars and their relation to durability, RILEM Seminar on the Durability of Concrete Structures Under Normal Outdoor Exposure, Hanover, March 1984.
3. Bakker, R. F. C., On the Cause of Increased Resistance of Concrete Made from Blast Furnace Cement to the Alkali Silica Reaction and to Sulphate Corrosion, D.Sc. thesis, Institut fur Gesteinstruttenkende, Aachen, 1980.
4. Regourd, M., Mortureux, B., and Gautier, E., Hydraulic reactivity of various pozzolanes, 5th Int. Symp. on Concrete Technology, Monterey, March 1981; Regourd, M., Mortureux, B., and Horman, H., Use of condensed silica fume as filler in blended cements, Proc. 1st Int. Conf. on the Use of Fly Ash, Silica Fume, Slag and Other Mineral Byproducts in Concrete, Montebello, Canada, July 1983, p. 847.
5. Huang, C.-Y. and Feldman, R. F., Influence of silica fume on the microstructure development of cement mortar, *Cement Concrete Res.,* 15(2), 285, 1985.
6. Huang, C.-Y. and Feldman, R. F., Hydration reactions in Portland cement-silica fume blends, *Cement Concrete Res.,* 15(4), 585, 1985.
7. Mikhail, R. Sh. and Youssef, A. M., Studies on fibre reinforced Portland cement pastes. I. Surface area and pore structure, *Cement Concrete Res.,* 4, 869, 1974.
8. Barnes, B. D., Diamond, S., and Dolch, W. L., Micromorphology of the interfacial zone around aggregates in Portland cement mortar, *J. Am. Ceram. Soc.,* 62(1-2), 21, 1979.
9. Takagi, N., Akashi, T., and Kakuta, S., Basic properties of Portland cement mortar containing condensed silica fume, *CAJ Rev.,* 116, 1983.
10. Feldman, R. F., Pore structure damage in blended cements caused by mercury intrusion, *J. Am. Ceram. Soc.,* 62(1), 30, 1984.
11. Feldman, R. F. and Huang, C.-Y., Properties of Portland cement silica fume pastes. I. Porosity and surface properties, *Cement Concrete Res.,* 15(5), 765, 1985.
12. Traetteberg, A., Frost action of blended cement with silica dust, in Durability of Building Materials and Components, ASTM STP691, American Society for Testing and Materials, Philadelphia, 1980, 536.
13. Delage, P. and Aitcin, P. C., Influence of Condensed Silica Fume on the Pore-Size Distribution of Concretes, I and E.C. Product Research and Development, Vol. 22, 1983, 286.
14. Huang, C.-Y. and Feldman, R. F., Dependence of frost resistance on the pore structure of mortar containing silica fume, *ACI J.,* 82(5), 740, 1985.
15. Buil, M., Paillere, A. M., and Roussel, B., High strength mortars containing condensed silica fume, *Cement Concrete Res.,* 14, 693, 1984.
16. Bager, D. H., Effect of silica fume on pore structure and drying shrinkage in mortars, *Norelisk Betong,* 3-4, 72, 1984.
17. Markestad, A. M., An Investigation of Concrete in Regard to Permeability Problems and Factors Influencing the Results of Permeability Tests, Rep. STF 65A 77027, Norwegian Institute of Technology, Trondheim, 1977.
18. Regourd, M., Mortureux, B., Aitcin, P. C., and Pinsonneault, P., Microstructure of field concretes containing silica fume, Proc. 4th Int. Conf. on Cement Microscopy, Las Vegas, 1982, 249.
19. Scheetz, B. E., Grutzek, M., Strickler, D. W., and Roy, D., Effect of composition of additives upon microstructures of hydrated Portland cement composites, Proc. 3rd Int. Conf. on Cement Microscopy, Houston, 1981, 307.
20. Carles-Giberques, A., Grandet, J., and Ollivier, J. P., in *Contact Zone Between Paste Aggregates,* Bartos, P., Ed., Applied Science, London, 1982, 24.
21. Bache, H. H., Densified cement — ultra fine particle based materials, 2nd Int. Conf. on Superplasticizers in Concrete, Ottawa, 1981.
22. Preece, C., Arup, H., and Frolund, T., Electro chemical behaviour of steel in dense silica cement, Proc. 1st Int. Conf. on the use of Fly Ash, Silica Fume and Slag and Other Mineral By-Products in Concrete, Montebello, Canada, 1983, 785.
23. Litvan, G. G., Air entrainment in the presence of superplasticizers, *ACI J.,* July-August, 326, 1983.
24. Mehta, P. K. and Gjorv, O. E., Properties of Portland cement concrete containing fly ash and condensed silica fume, *Cement Concrete Res.,* 12, 587, 1982.
25. Carette, G. G. and Malhotra, V. M., Silica Fume in Concrete — Preliminary Investigation, CANMET Report M-38-82-IE, 1982, 1.

26. Virtaven, J., Freeze-thaw resistance of concrete containing blast-furnace slag, fly-ash or condensed silica fume, Proc. 1st Int. Conf. on the Use of Fly Ash, Silica Fume, Slag and Other Mineral By-Products in Concrete, Montebello, Canada, 1983, 923.

27. Sorensen, E. V., Freezing and thawing resistance of condensed silica fume concrete exposed to deicing chemicals, Proc. 1st Int. Conf. on the Use of Fly Ash, Silica Fume and Other Mineral By-Products in Concrete, Montebello, Canada, 1983, 709.

27a. Hooton, D., Private communication.

28. Feldman, R. F., The effect of sand cement ratio and silica fume on the microstructure and freeze-thaw durability of mortars, Proc. 2nd Int. Conf. on Use of Fly Ash, Silica Fume, Slag, and Natural Pozzolans in Concrete, 2, 973, 1986.

29. Popovic, K., Ukraincik, V., and Djurekovic, A., Improvement of mortar and concrete durability by the use of condensed silica fume, *Durability Build. Mater.*, 2, 171, 1984.

30. Feldman, R. F. and Huang, C.-Y., Resistance of mortars containing silica fume to attack by a solution containing chlorides, *Cement Concrete Res.*, 15(3), 411, 1985.

31. Mather, K., Factors affecting sulphate resistance of mortars, 7th Int. Congr. on the Chemistry of Cement, Vol. 4, Paris, 1980, 580.

32. Oberholster, R. E. and Westra, P., The effectiveness of mineral admixtures in reducing expansion due to alkali reaction with Malmesbury group aggregates, in Alkali-Aggregate Reaction in Concrete, Conf. in Cape Town, 1981, 294.

33. Olafson, H., The Effect of Finely Grained Silica Dust and Fly Ash on Alkali Silica Reactivity on High Alkali Cements, National Bureau of Standards Building Composite Group, Washington, D.C., 1980, 1.

34. Brotschi, J. and Mehta, P. K., Test methods for determining potential alkali silica reactivity of cements, in press.

35. Perry, C. and Gillott, J. E., Possibility of using silica fume to control expansion to alkali silica reaction, in press.

36. Diamond, S., Effects of microsilica (silica fume) on pore solution chemistry of cement pastes, *Commun. Am. Ceram. Soc.*, C-82-84, 1983.

37. Grattan-Bellew, P. E., unpublished data.

Chapter 7

PROPERTIES OF FRESH CONCRETE

I. INTRODUCTION

The properties of fresh concrete are primarily governed by the mixture proportioning criterion and the mixture ingredients. When a mineral admixture is added to a basic concrete mixture it can affect its rheological behavior, water demand, bleeding, plastic shrinkage, and setting time characteristics. The change in some or all of these properties will, in turn, affect the consolidation effort needed to properly compact concrete in place, and the long-term properties of hardened concrete such as strength, volume stability, permeability, and durability to aggressive agents. This chapter will discuss the effect of condensed silica fume on the properties of fresh concrete.

II. COLOR

Condensed silica fume varies considerably in color depending upon its source, i.e., whether it is from a ferrosilicon or silicon furnace and also upon the amount of carbon it may contain.

The fresh concrete incorporating condensed silica fume is generally darker gray in color than fresh conventional concrete. This is particularly so for concrete incorporating higher percentages of condensed silica fume with a higher percentage of carbon.

In one investigation performed in Sweden, the degree of color was determined according to a Swedish standard.[1] The concretes were made using a light and a dark colored fume. The use of dark condensed silica fume resulted in fresh concrete having a darker appearance than the control concrete, and that containing a light colored condensed silica fume. However, when the test specimens made from these concretes were stored in a laboratory environment for 20 days, there was no significant difference in color between the two types of concrete. No definite explanation can be offered for this except that the initial difference in color was neutralized by drying and perhaps by carbonation. However, in actual structural members, silica fume concrete appears to be somewhat darker.

III. WATER DEMAND

Condensed silica fume, because of its high surface area, has great affinity for water and this is reflected in concrete incorporating the fume. For example, in concrete made with a water-to-cement ratio of 0.64, the increase in water demand is almost directly proportional to the amount of added fume[2] (Figure 1). The effects of silica fume and cement content on the water demand of concretes containing two dosages of plasticizers are compared in Figure 2.[3] The water demand decreases at all dosages of silica fume because of the dispersing action of the plasticizers. Further reduction in the water demand is achieved by the incorporation of a superplasticizer which is a better dispersant than the normal plasticizer. The effect of plasticizers on the water demand is evident, independent of the type of cement and its amount used in concrete. Another important observation is, although the water demand increases with the amount of silica fume in reference concrete, that the water demand is only marginally affected in concretes containing the plasticizers.

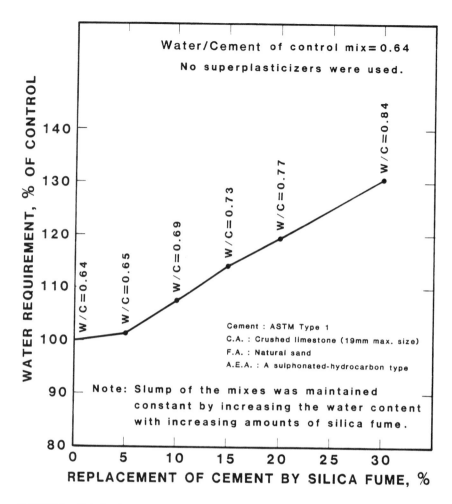

FIGURE 1.   Relation between water requirement and dosage of silica fume for concrete with a W/(C + SF) of 0.64.

The water demand increase per kilogram of condensed silica fume added may be of the order of 1 ℓ.[4] However, Loland and Hustad[5] and Aitcin et al.[6] have reported that for lean concretes (cement content 100 to 150 kg/m³), the water demand decreases upon the addition of condensed silica fume (Figure 3). Comparison of water demand values at cement contents of 100 kg/m³ and 250 kg/m³ indicates that in all mixes containing the plasticizer and silica fume water demand is decreased. In the mix containing 10/0 (silica fume/plasticizer), the water demand is slightly lower in the concrete containing 100 kg/m³ of cement. The reduced water demand incorporating condensed silica fume is of academic interest only because durability requirements, at least in North America, do not allow the use of concretes with such low cement content in structural concrete.

The increased water demand of concrete incorporating condensed silica fume can be overcome with the use of water reducers and superplasticizers (high-range water reducers), and to a lesser degree by reducing the fine aggregate content of a mix. The original investigations in Norway were performed using lignosulfonate-based water reducers, and the Norwegian data show that with a lignosulfonate-based water reducer dosage of 0.20 to 0.40% by weight of cement, the water demand of the control concrete

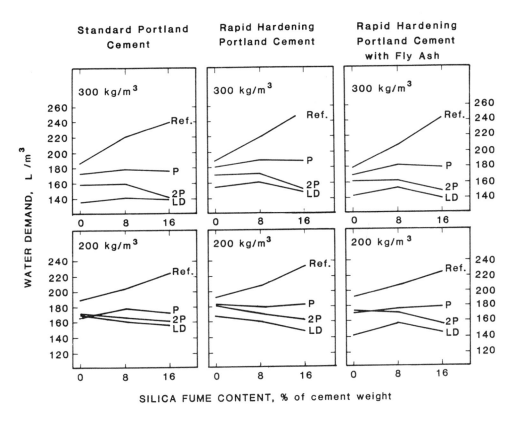

FIGURE 2. Relation between water demand and dosage of silica fume for concrete made with different types of cements.

and that incorporating 10% condensed silica fume becomes equal. It has been claimed[4] that lignosulfonate-based water reducers are as efficient or even more so in reducing water demand in condensed silica fume concrete than superplasticizers. However, no substantial published data are available to support this claim. The use of high dosage of lignosulfonate-based water reducers can cause serious delays in the setting of fresh concrete. The use of high-range water reducers is preferable unless laboratory investigations dictate otherwise.

IV. AIR ENTRAINMENT

The dosage of an air-entraining admixture to produce a required volume of air in concrete increases markedly with increasing amounts of condensed silica fume (Figure 4). The increased demand for an air-entraining admixture is probably due to the extremely high surface area of the fume, especially if it also contains carbon.[2] Similar results have been reported when fine filler materials such as crusher dust are incorporated into concrete.[7] The retention of air content in fresh concrete is generally not affected. The published data indicate that in low (W/C) ratio concretes containing more than 20% condensed silica fume, it is difficult to entrain more than 4 to 5% air.[2] In the investigation referred to above, a sulfonated hydrocarbon type of air-entraining admixture was used. Whether a vinsol resin type air-entraining admixture would have given similar results is not known.

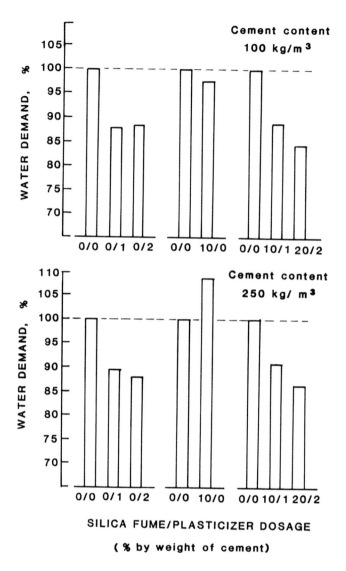

FIGURE 3.   Relation between water demand and silica fume/plasticizer dosage rate.

V. BLEEDING

Concrete incorporating condensed silica fume shows reduced bleeding because of the change in its rheological properties. These changes are expected because condensed silica fume has high affinity for water resulting in very little water left in the mix for bleeding. Silica fume particles attach themselves to adjacent cement particles and reduce the channels for bleeding.

Loland and Hustad[5] quantitatively evaluated the bleeding and segregation aspects of a large number of concrete mixes. They concluded that bleeding was greatly reduced in concrete incorporating condensed silica fume. In their investigation of the bleeding of fresh concrete, Johansén[8] and Maage[9] also found that in concrete with condensed silica fume as a cement replacement, a large reduction in bleeding resulted. This was true for condensed silica fume concrete mixtures with and without water reducing admixtures.

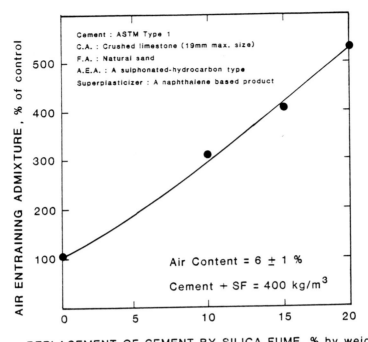

FIGURE 4.   Relation between air entraining admixture demand and dosage of silica fume.

Data[10] on bleeding of concrete incorporating various percentages of condensed silica fume are shown in Table 1 and Figure 5, and they indicate significant reduction in bleeding. The bleeding test was done according to ASTM C232.[11] Similar data for lightweight concrete have been published by Bürge[12] and are shown in Figure 6.

VI. COHESIVENESS

Concrete containing condensed silica fume appears more cohesive and less prone to segregation. Investigations performed in Norway[4] have shown that at equal slumps, concrete containing condensed silica fume requires more energy input for a given flow in a "Thalow Stroke" test which is somewhat similar to the ASTM Flow Table Test.[13] Also, there appears to be a tendency for the condensed silica fume concrete to lose slump more rapidly than the control concrete. Furthermore, concrete incorporating more than 10% condensed silica fume becomes more sticky and "gluey" and increased vibration is needed to make it flow. This may perhaps be due to the silica fume imbibing water to form a gel. In order to overcome the above problem and to maintain the same consistency for some length of time, it is necessary to increase the initial slump of the condensed silica fume concrete by about 50 mm. On the other hand, the increased cohesiveness of the mixture is a decided advantage in "flowing" concrete and pumped concrete.

VII. SETTING TIME

The setting time of Portland cement concrete in North America is normally measured by sieving mortar from a concrete mixture and determining its resistance to pene-

Table 1

MIX PROPORTION, PROPERTIES OF FRESH CONCRETE, AND BLEEDING RATES[10]

			Mix proportions										Properties of fresh concrete										
Mix no.	W/(C+SF) (by wt)	Cement (kg/m³)	Silica fume %	Silica fume kg/m³	Fine agg. (kg/m³)	Coarse agg. (kg/m³)	A.E.A. (ml)	SP[a] (kg)	T (°C)	Slump (mm)	Unit wt (kg/m³)	Air content (%)	Bleeding,[b] cm³/cm² at (min)								
													10	20	30	40	70	130	170	200	300
1	0.40	412	0	0	667	1090	262	0	22	75	2330	5.5	Nil	—	—	—	—	Nil	—	—	Nil
2	0.40	381	5	21	653	1063	644	2.0	22	90	2280	7.2	Nil	—	—	—	—	Nil	—	—	Nil
3	0.40	368	10	40	659	1074	1071	4.0	22	95	2305	6.2	Nil	—	—	—	—	Nil	—	—	Nil
4	0.40	346	15	61	655	1067	2246	5.5	21	115	2295	6.4	Nil	—	—	—	—	Nil	—	—	Nil
5	0.64	263	0	0	744	1213	136	0	22	90	2390	3.0	0.003	0.013	0.023	0.032	0.069	0.141	—	0.157	0.157
6	0.64	246	5	12	726	1182	178	0	21	85	2332	4.5	0	0.002	0.005	0.010	0.032	0.070	0.081	0.082	0.082
7	0.64	229	10	24	709	1156	262	0	21	75	2281	6.0	0	0	0	0	0.003	0.009	0.012	0.012	0.012
8	0.64	221	15	39	724	1182	581	2.0	21	75	2332	4.8	0	0	0	0	0	0	0	0	0

[a] A naphathalene-based superplasticizer was used.
[b] ASTM C232.

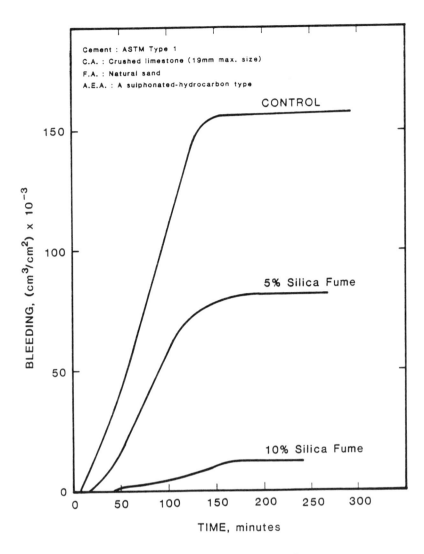

Cement : ASTM Type 1
C.A. : Crushed limestone (19mm max. size)
F.A. : Natural sand
A.E.A. : A sulphonated-hydrocarbon type

CONTROL

5% Silica Fume

10% Silica Fume

BLEEDING, $(cm^3/cm^2) \times 10^{-3}$

TIME, minutes

FIGURE 5.   Bleeding rate of control and silica fume concrete.

tration by needles of a given bearing area. The concrete is said to have reached initial set when the mortar reaches a penetration resistance of 3.5 MPa and final set when it reaches a penetration resistance of 27.6 MPa.[14] Data on the setting time of concrete incorporating condensed silica fume are sparse. This is partly because condensed silica fume concrete usually incorporates water reducing admixtures and/or superplasticizers, and the use of the latter tends to mask the effect of condensed silica fume. Limited investigations[10] show that the setting time is not significantly affected by the incorporation of condensed silica fume. Figures 7 and 8 show setting times of nonsuperplasticized and superplasticized condensed silica fume concretes, respectively. Notwithstanding the fact that the use of a superplasticizer may affect the setting time of concrete, the data in Table 1 and Figures 7 and 8 do bring out the negligible effect of the use of 5 and 10% condensed silica fume on the setting time of concrete. However, for concrete with a W/(C + SF) of 0.40 and 15% condensed silica fume (Figure 8), there is a marked increase in the setting time. This is primarily due to the high dosage of the superplasticizer used.

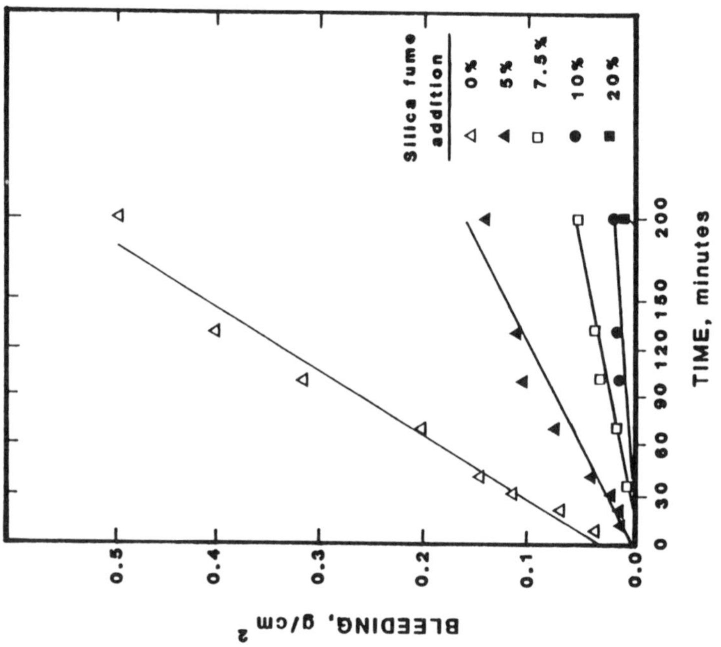

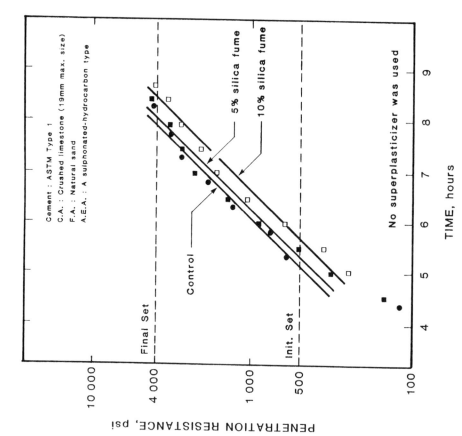

FIGURE 7. Penetration resistance of control and nonsuperplasticized silica fume concrete.

FIGURE 6. Bleeding rate of lightweight silica fume concrete.

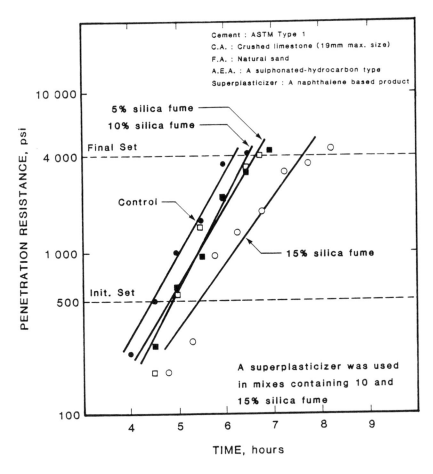

FIGURE 8. Penetration resistance of control and superplasticized silica fume concrete.

VIII. PLASTIC SHRINKAGE

Plastic shrinkage cracking of fresh concrete is associated with the curing conditions which allow a faster rate of evaporation of water from the surface of fresh concrete than the water replaced by bleeding from the concrete underneath. Thus, all chemical and mineral admixtures which reduce bleeding of fresh concrete make it more prone to plastic shrinkage cracks. This is particularly so for condensed silica fume concrete in which bleeding is considerably reduced. This problem of plastic shrinkage cracks can become very serious under curing conditions of high temperature and high wind velocity which favor faster evaporation of water from freshly placed concrete. According to the investigations reported from Norway, the most vulnerable period for plastic shrinkage cracks to appear is the time when concrete is about to set.[4]

REFERENCES

1. Bellander, V., Experiences with Silica Fume in Ready-Mixed Concrete, Rep. 82.610, Division of Building Materials, The Norwegian Institute of Technology, Trondheim, 1982.
2. Carette, G. G. and Malhotra, V. M., Mechanical Properties, Durability and Drying Shrinkage of Portland Cement Concrete Incorporating Silica Fume; *ASTM J. Cement, Concrete, Aggregates,* 5(1), 3, 1983.
3. Sellevold, E. J. and Radjy, F. F., Condensed Silica Fume (Microsilica) in Concrete: Water Demand and Strength Development, ACI Spec. Publ. SP-79, Malhotra, V. M., Ed., American Concrete Institute, Detroit, 1983, 677.
4. Sellevold, E. J., Review: Microsilica in Concrete, report no. 08037/EJS/TJJ, Norwegian Building Research Institute, Oslo, 1984.
5. Loland, K. E. and Hustad, T., Report 1: Fresh Concrete and Methods of Data Analysis, rep. no. STF65 A81031, Cement and Concrete Research Institute, The Norwegian Institute of Technology, Trondheim, June 1981.
6. Aitcin, P. C., Pinsonneault, P., and Rau, G., in *Proceedings, The Materials Society Symposium on Fly Ash,* Diamond, S., Ed., Boston, 1981, 316.
7. Malhotra, V. M. and Carette, G. G., Performance of concrete incorporating limestone dust as partial replacement for sand, *ACI Proc.,* 81(3), 363, 1985.
8. Johansén, R., Silicastov i fabrickksbetong Langtidseffekter, rep. no. STF65 F79019, Cement and Concrete Research Institute, The Norwegian Institute of Technology, Trondheim, May 1979.
9. Maage, M., Modifisert Portland Cement, rep. no. STF65 A83063, Cement and Concrete Research Institute, The Norwegian Institute of Technology, Trondheim, October 1983.
10. Bilodeau, A., Influence des Fumées de Silice sur le Ressuage et le Temps de Prise du Béton, CANMET rep. no. MRP/MSL 85-22 (TR), 1985.
11. ASTM C 232-71 (1977), Standard test method for bleeding of concrete, ASTM 1984 Annual Book of Standards, Vol. 04.02, Sect. 4, American Society for Testing and Materials, Philadelphia, 1984.
12. Bürge, A. T., High-Strength Lightweight Concrete with Condensed Silica Fume, ACI Spec. Publ. SP-79, Malhotra, V. M., Ed., American Concrete Institute, Detroit, 1983, 731.
13. ASTM C230-83, Standard specification for flow table for use in tests of hydraulic cements, ASTM 1984 Annual Book of Standards, Vol. 04.02, Sect. 4, American Society for Testing and Materials, Philadelphia, 1984.
14. ASTM C403-80, Standard test method for time-of-set of concrete mixtures by penetration resistance, ASTM 1984 Annual Book of Standards, Vol. 04.02, Sect. 4, American Society for Testing and Materials, Philadelphia, 1984.

Chapter 8

PROPERTIES OF HARDENED CONCRETE

I. INTRODUCTION

The properties of hardened concrete involve primarily its strength and modulus of elasticity; other properties required often are the drying shrinkage, creep, permeability, bond, and durability. In general, strength properties of concrete are the most important because they are related to the structure of hardened cement paste, are relatively easy to determine, and can be used to estimate its other properties. However, there are instances when durability and permeability of concrete may become the governing criteria, e.g., in parking structures and highway bridge decks. This chapter presents data on mechanical properties and permeability of condensed silica fume concrete; the durability aspects are covered in Chapter 9.

II. COMPRESSIVE STRENGTH

The strength development characteristics of condensed silica fume concrete are similar to those of fly ash concrete except that the results of the pozzolanic reactions of the former are evident at early ages. This is due to the fact that condensed silica fume is a very fine material with a very high glass and silica content. Figure 1 shows schematically the filler and pozzolanic effects of silica fume in concrete.[1] The overall effect of the various factors would indicate the requirement of an optimum amount of silica fume addition. This is because the upward trend of pozzolanic and microparticle effects are countered by the increased water demand as the dosage of silica fume is increased.

Figure 2 gives strength results for concrete when condensed silica fume (SF) is used as a direct replacement (by weight) for Portland cement,[2] and Figure 3 refers to concrete when condensed silica fume is used as an additive.[3] When silica fume is used as a direct replacement for Portland cement, there is little or no change in compressive strength at early ages (1 to 3 days) for concrete with high water-to-(cement + silica fume) (W/(C + SF)) ratios.[4] This is regardless of the percentage of the fume used. This would indicate that the pozzolanic effect of silica fume requires some minimum amount of $Ca(OH)_2$ formation. However, at lower W/(C + SF) and higher percentages of incorporated fume, there is a marked increase in strength at 3 days (Figure 4). At lower W/(C + SF), the products formed by the pozzolanic reaction aggregate particles more efficiently than when they are formed at higher W/(C + SF). This effect is noticed both in air-entrained and nonair-entrained concrete.

Investigators in Norway have performed considerable research on condensed silica fume to develop data on its relative strength-producing property in comparison with that of Portland cement. This property has been variously called the activity index, efficiency index, or substitution index. According to Jahren,[1] the value of the above factor ranges from 2 to 5. For example, the efficiency index of silica fume is said to be 3, if 3 kg of cement can be replaced by 1 kg of silica fume and the resulting strength is the same as that of control concrete. Figure 5 shows the relationship between the percentage of silica fume by weight of cement and the efficiency index. It would appear that the percentage increase in strength of concrete is higher, the lower the silica fume content. The relationship between W/C and the 28-day compressive strength of concrete containing different percentages of silica fume addition is shown in Figure 6.[5]

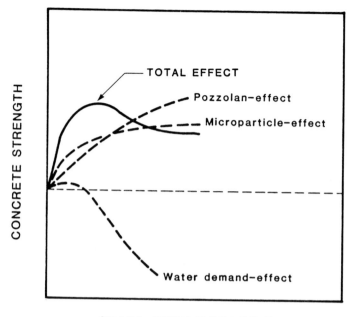

FIGURE 1.   A generalized relationship between silica fume dosage rate and concrete strength.

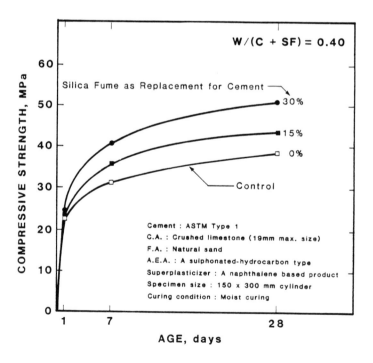

FIGURE 2.   Relation between compressive strength and age for concrete incorporating various percentages of silica fume as partial replacement for cement.

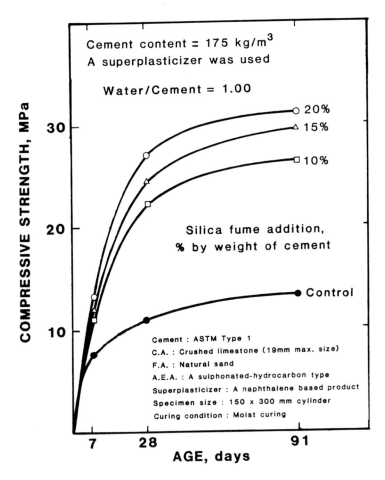

FIGURE 3.   Relation between compressive strength and age for concrete incorporating various percentages of silica fume as an addition to cement.

These data cover a wide range of mix proportions, curing conditions, and testing ages. Some mixes contained a water reducer and a superplasticizer. It is clear from the data that the efficiency of silica fume in concrete increases with decreasing W/C ratio.

As mentioned already, the water demand of silica fume concrete is directly proportional to the amount of silica fume (used as a percentage replacement of Portland cement) if the slump of concrete is maintained constant by increasing the water content rather than by using a superplasticizer. In such instances, the increase in the strength of silica fume concrete over that of the control concrete is only observed at lower W/C ratios and at longer ages. One such case is shown in Figure 7 which illustrates the relationship between compressive strength expressed as a percent of control strength, and the percent replacement of cement by condensed silica fume.[2] Figure 8 shows the same data except the 91-day strength test results in a more conventional manner, i.e., compressive strength vs. age.

Tables 1 to 3 provide strength test data from a ready-mixed concrete producer on silica fume concrete for different maximum size aggregates.[6] The compressive strength values at 7, 14, and 28 days, together with the percentage of silica fume used and slump of fresh concrete, are shown in the tables. It is obvious that the so-called efficiency factor increases with increasing cement content of concrete mixtures.

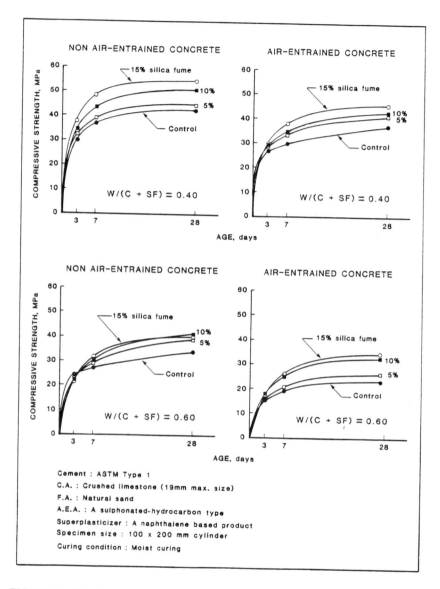

FIGURE 4. Relation between compressive strength and age for silica fume concrete made with W/(C + SF) of 0.40 and 0.60.

III. FLEXURAL/TENSILE STRENGTH

A generalized relationship between the ratio of flexural or tensile strength to compressive strength and the silica fume replacement of cement in concrete is shown in Figure 9.[1] The large variations in results occur because data were compiled from tests performed in different laboratories which had different test procedures and size of test specimens.

The ratio of the 14-day flexural strength to the 28-day compressive strength of silica fume concrete follows the same pattern as for normal weight concrete (Tables 4 and 5). For example, at a compressive strength level of about 35 MPa, the ratio is of the order of 20%; at a compressive strength level of 55 MPa, the ratio is of the order of 10%.[4] Additional data on compressive, flexural, and splitting strengths are shown in Table 6.[3]

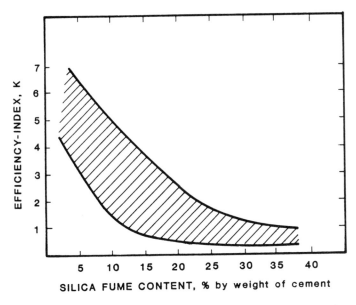

FIGURE 5. A generalized relation between efficiency index and silica fume content of concrete.

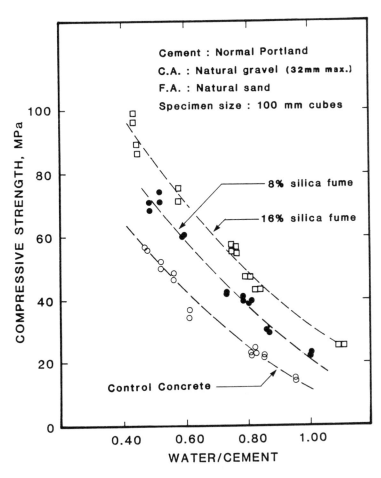

FIGURE 6. Relation between compressive strength of silica fume concrete and W/C ratio.

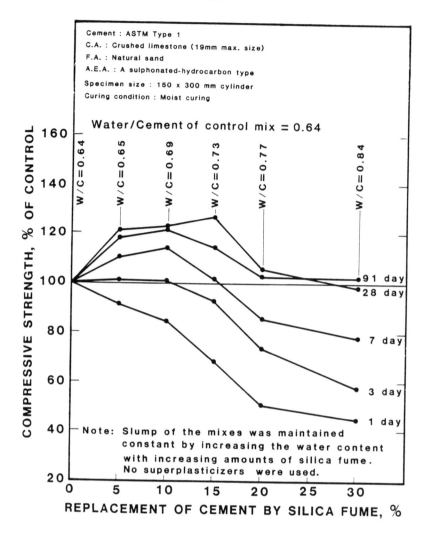

FIGURE 7.   Relation between compressive strength of concrete and dosage of silica fume (W/C) = 0.64.

IV. YOUNG'S MODULUS OF ELASTICITY

Comprehensive investigations in Norway[7] have shown that there are no significant differences between the Young's modulus of elasticity "E" of concrete with and without condensed silica fume. However, it is known that "E" values do not continue to increase with increasing compressive strength of concrete. Therefore, very high-strength concretes tend to be more brittle. This is equally true of high-strength condensed silica fume concrete. Malhotra et al.[8] have reported data on the Young's modulus of elasticity of Portland cement/blast-furnace slag/silica fume concrete (Table 7). They found that regardless of the various percentages of silica fume contents and W/(C + BFS) ratios there is no significant difference between the "E" values obtained at 28 days. There was some indication that the highest values were obtained for concrete containing 10% silica fume.

Wolsiefer[9] reported a Young's modulus value of 43.1 GPa for condensed silica fume concrete having a 28-day compressive strength of 97.9 MPa. The mixture proportions of this concrete are shown in Table 8.

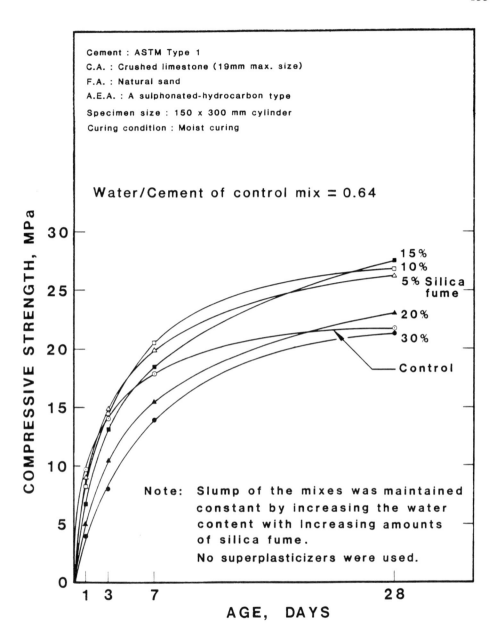

FIGURE 8. Relation between compressive strength and age for silica fume concrete: W/C = 0.64.

V. CREEP

No published data are available on creep of concrete incorporating 5 to 15% of condensed silica fume. However, Wolsiefer[9] has reported limited data on concrete incorporating a propriety admixture (corrocem) which is principally silica fume (about 90%). Creep measurements of the high-strength concrete incorporating the above admixture were made after 7 days and every month after that up to 1 year. The applied load was 17 MPa which was applied at 12 hr and 28 days, and the test was conducted in accordance with ASTM C512. Data on reference mixes have not been reported.

Table 1
RESULTS OF COMPRESSIVE STRENGTH TESTS AT 7, 14, AND 28 DAYS FOR CONCRETES MADE WITH AND WITHOUT CONDENSED SILICA FUME — 10 mm MSA[6]

10 mm MSA (in brackets: control mixes)

	20 MPa		25 MPa		30 MPa		35 MPa	
Cement (kg/m³)	200	(268)	227	(310)	255	(360)	273	(400)
Silica fume (kg/m³)	24.0	—	27.2	—	28.1	—	27.3	—
Silica fume (%)	12.0	—	12.0	—	11.0	—	10.0	—
Compressive 7 days	13.4	(15.8)	17.8	(19.8)	21.0	(24.7)	24.5	(27.2)
strength 14 days	18.1	(19.2)	23.6	(23.5)	27.5	(28.5)	32.8	(33.4)
(MPa) 28 days	23.5	(22.0)	30.4	(26.4)	35.7	(33.2)	41.4	(37.2)
Slump (mm)	75—125		70—125		65—110		65—110	
Efficiency factor	2.9		3.5		4.3		5.3	

Note: MSA — Maximum size aggregates; specimen size = 150 × 300-mm cylinders.

Table 2
RESULTS OF COMPRESSIVE STRENGTH TESTS AT 7, 14, AND 28 DAYS FOR CONCRETES MADE WITH AND WITHOUT CONDENSED SILICA FUME — 14 mm MSA[6]

14 mm MSA (in brackets: control mixes)

	20 MPa		25 MPa		30 MPa		35 MPa	
Cement (kg/m³)	186	(250)	200	(282)	245	(341)	273	(386)
Silica fume (kg/m³)	22.3	—	24.0	—	27.0	—	27.3	—
Silica fume (%)	12.0	—	12.0	—	11.0	—	10.0	—
Compressive 7 days	14.1	(16.6)	17.9	(18.9)	23.1	(26.8)	28.1	(27.8)
strength 14 days	18.1	(17.4)	20.6	(23.0)	31.9	(33.2)	34.1	(32.6)
(MPa) 28 days	24.6	(22.0)	27.9	(26.0)	36.4	(34.0)	40.8	(36.8)
Slump (mm)	75—125		80—125		70—130		65—110	
Efficiency factor	3.5		3.7		4.2		5.2	

Note: Specimen size = 150 × 300-mm cylinders.

The creep results are given in Table 8. The very low creep values are for the specialized concrete used in the test program, and should not be used for comparison purposes for concretes incorporating lower percentages of silica fume, and with compressive strengths of the order of 40 to 60 MPa at 28 days.

VI. DRYING SHRINKAGE

The drying shrinkage of concrete is controlled by the volume fraction of the cement paste, and volume fraction, stiffness, and maximum size of the aggregate. Also, the conditioning of the test specimens before the beginning of drying shrinkage seriously affects the test results. Several investigators have reported data on unrestrained shrinkage of standard specimens at a relative humidity of 50 to 60%, but test results are generally difficult to compare because of the different mix proportions and different curing periods used before drying shrinkage measurements.[10,11]

Table 3
RESULTS OF COMPRESSIVE STRENGTH TESTS AT 7, 14, AND 28 DAYS FOR CONCRETES MADE WITH AND WITHOUT CONDENSED SILICA FUME — 20 mm MSA[6]

20 mm MSA (in brackets: control mixes)

	20 MPa		25 MPa		30 MPa		35 MPa	
Cement (kg/m³)	182	(245)	195	(277)	240	(336)	273	(395)
Silica fume (kg/m³)	21.8	—	21.5	—	26.4	—	27.3	—
Silica fume (%)	12.0	—	11.0	—	11.0	—	10.0	—
Compressive 7 days	14.9	(18.0)	18.6	(22.0)	23.4	(27.1)	27.1	(31.0)
strength 14 days	19.2	(20.4)	25.1	(26.6)	27.5	(30.0)	33.8	(35.2)
(MPa) 28 days	26.7	(24.1)	33.2	(29.5)	34.6	(34.5)	40.2	(37.6)
Slump (mm)	80—135		60—130		65—120		60—105	
Efficiency factor	3.6		3.9		4.1		5.0	

Note: Specimen size = 150 × 300-mm cylinders.

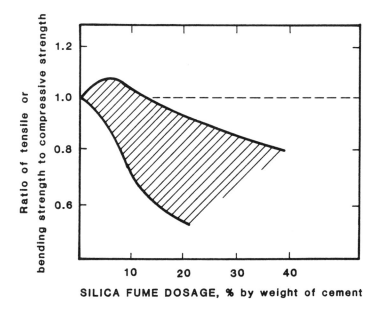

FIGURE 9.   A generalized relation between ratio of tensile or bending strength to compressive strength and silica fume dosage rate.

Johansen[10] performed shrinkage measurements on concrete prisms that were exposed to a drying environment (50% relative humidity) immediately after demolding, and after 28-day moist curing. The condensed silica fume content ranged from 0 to 25% and W/(C + SF) ratio ranged from 0.37 to 1.06. Some mixes contained water reducers. It was concluded that for concrete with W/(C + SF) < 0.60, no significant differences in shrinkage existed between the reference concrete and condensed silica fume concrete containing up to 10% silica fume. Concrete containing 25% silica fume and no water reducers showed higher shrinkage values.

Loland and Hustad[11] have reported shrinkage data on condensed silica fume con-

Table 4

SUMMARY OF COMPRESSIVE AND FLEXURAL STRENGTHS: NONAIR-ENTRAINED CONCRETE[4]

Mix no.	W/(C + SF) (by wt)	Cement (kg/m³)	Silica fume (kg/m³)	Density at 1 day (kg/m³)		Compressive strength[a] of 102 × 203 mm cylinders at various ages (MPa)			Flexural strength[a] of 76 × 102 × 406 mm prisms (MPa)
				76 × 102 × 406 mm prisms	102 × 203 mm cylinders	3-day	7-day	28-day	14-day
1	0.60	284	0	2249	2446	24.0	27.5	34.1	6.0
2	0.60	267	14	2413	2412	23.2	28.6	39.2	6.3
3	0.60	254	29	2404	2406	23.1	30.3	41.6	6.2
4	0.60	239	43	2389	2400	21.6	31.1	39.4	6.8
5	0.50	340	0	2444	2442	28.0	32.5	38.3	5.7
6	0.50	323	17	2420	2419	28.0	34.1	40.6	6.4
7	0.50	305	34	2410	2419	29.5	38.5	49.0	7.2
8	0.50	289	51	2398	2406	31.9	41.1	50.4	7.6
9	0.40	431	0	2436	2440	30.6	37.2	42.6	6.2
10	0.40	415	22	2448	2429	32.4	39.4	44.2	7.7
11	0.40	389	43	2416	2406	33.2	43.6	50.2	7.8
12	0.40	367	65	2409	2398	37.1	49.0	54.4	8.0
13	0.40	302	130	2387	2392	40.9	51.6	67.7	7.1

[a] Each compression test result is the mean of tests on three cylinders and each flexural strength test result is the mean of tests on two prisms.

Table 5

SUMMARY OF COMPRESSIVE AND FLEXURAL STRENGTHS: AIR-ENTRAINED CONCRETE[a]

| Mix no. | W/(C + SF) (by wt) | Cement (kg/m³) | Silica fume (kg/m³) | Density at 1 day (kg/m³) | | Compressive strength[a] of 102 × 203 mm cylinders at various ages (MPa) | | | Flexural strength[a] of 76 × 102 × 406 mm prisms (MPa) |
				76 × 102 × 406 mm prisms	102 × 203 mm cylinders	3-day	7-day	28-day	14-day
1	0.60	243	0	2299	2346	15.8	18.9	23.1	4.4
2	0.60	224	11	2236	2272	15.8	20.2	26.2	4.3
3	0.60	221	23	2274	2317	17.9	25.5	32.5	5.1
4	0.60	202	36	2268	2306	17.1	25.7	34.0	5.1
5	0.40	417	0	2345	2349	26.9	29.9	36.9	6.5
6	0.40	393	21	2330	2347	27.8	33.7	41.3	7.2
7	0.40	364	40	2290	2292	27.9	34.4	41.9	7.1
8	0.40	340	61	2282	2280	27.8	38.0	45.2	7.0
9	0.42	288	30	2300	2319	32.6	45.2	57.0	6.0

[a] Each compression test result is the mean of tests on three cylinders and each flexural strength test result is the mean of tests on two prisms.

Table 6
COMPRESSIVE, FLEXURAL, AND SPLITTING-TENSILE TEST RESULTS AT VARIOUS AGES[3]

Mix	Type	W/(C + SF)[a]	Compressive strength of 102 × 203 mm cylinders (MPa)					Flexural strength of 76 × 102 × 406 mm prisms (MPa)		Splitting-tensile strength of 102 × 203 mm cylinders (MPa)	
			1-day	3-day	7-day	28-day	91-day	14-Day	28-Day	7-day	28-day
1	Control	0.64	9.8	14.1	18.0	21.7	24.8	4.5	4.6	2.7	3.2
2	5% Silica fume	0.65	9.0	14.3	19.9	26.4	29.3	5.3	5.4	2.7	3.4
3	10% Silica fume	0.69	8.3	14.2	20.6	26.7	30.5	4.7	5.2	2.8	3.6
4	15% Silica fume	0.73	6.7	13.2	18.3	27.6	28.4	4.5	5.0	2.8	3.3
5	20% Silica fume	0.77	5.0	10.4	15.5	23.0	25.4	4.3	4.5	2.5	3.3
6	30% Silica fume	0.84	4.4	8.1	14.1	21.4	25.3	3.7	4.5	2.2	2.8

Note: Each compressive and splitting-tensile strength value is average of three tests. Each flexural strength value is average of two tests.

[a] Water/(cement + silica fume) (by weight).

Table 7
COMPRESSIVE STRENGTH AND MODULUS OF ELASTICITY
OF CONCRETE WITH DIFFERENT
W/(C + BFS)[8]

W/(C + BFS)[a]	Mix no.	Type of mixture[b]	Compressive strength of 100 × 200-mm cylinders (MPa)		Modulus of elasticity (GPa) 28 days
			7-day	28-day	
	1	Reference	38.3	47.1	36.0
	2	Control	25.7	46.7	36.3
0.40	3	5% Silica fume	30.3	57.1	36.1
	4	10%	34.3	59.3	36.9
	5	15%	34.5	60.7	35.8
	6	20%	37.6	60.1	34.9
	7	Reference	30.4	36.5	33.3
	8	Control	18.8	33.0	34.1
0.50	9	5% Silica fume	20.7	38.9	33.6
	10	10%	25.1	45.5	34.4
	11	15%	28.5	48.8	34.0
	12	20%	29.3	47.5	33.8
	13	Reference	21.0	28.3	31.3
	14	Control	10.0	20.6	32.6
0.65	15	5% Silica fume	11.9	26.2	31.4
	16	10%	15.7	31.7	32.8
	17	15%	18.5	35.3	31.8
	18	20%	22.7	38.0	31.3

Note: Each value is average of three tests.

[a] Water/(cement + blast furnace slag) by weight = W/(C + BFS).

[b] Reference mixture: 100% normal Portland cement; control mixture: 50% normal Portland cement plus 50% BFS; silica fume mixture: 50% normal Portland cement plus 50% BFS plus additions of silica fume.

Table 8
MIX PROPORTIONS SHRINKAGE AND CREEP OF HIGH-
STRENGTH CONDENSED SILICA FUME CONCRETE[9]

Mix proportions (kg/m³)			Shrinkage and creep values[a]		
					Creep
		Time	Shrinkage 10^{-6}	$\times 10^{-6}$	10^{-6}/kPa
Type I Portland cement	= 593	7 day	60	120	0.007
Silica fume admixture	= 119	28 day	120	210	0.012
F.A.	= 537	3 months	170	330	0.019
C.A.	= 997	6 months	270	425	0.025
Water	= 158	12 months	315	480	0.028
W/(C + SF)	= 0.22				
Average slump	= 119 mm				
Average air content	= 1.5%				

[a] Creep load = 17.2 MPa; loading age = 28 days; compressive strength at loading = 111.4 MPa; stress/strength ratio = 17.2/111.4 = 0.154.

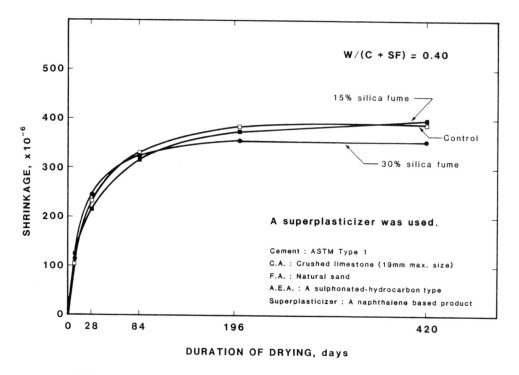

FIGURE 10.   Relation between shrinkage and duration of drying for concrete: W/(C + SF) = 0.40.

crete with varying W/(C + SF) and containing up to 20% silica fume. The conditioning of test specimens consisted of 7 days of moist curing followed by drying at 60% relative humidity. The general conclusion was that the shrinkage of the condensed silica fume concrete was comparable to that of the reference concrete.

Data by Carette and Malhotra[2] indicate that the drying shrinkage of condensed silica fume concrete after 28 days of moist curing is generally comparable to that of the control concrete regardless of the W/C materials ratio. The tests were performed in accordance with ASTM C157 which stipulates storage of test specimens at 23 ± 1.1°C and 50 ± 4% relative humidity. Figure 10 shows drying shrinkage of concrete with a W/(C + SF) of 0.40; the coarse and fine aggregates were limestone and natural sand and the concrete was made using a naphthalene-based superplasticizer. The drying shrinkage of control concrete and that incorporating 15% silica fume are comparable while concrete containing 30% silica fume shows slightly lower shrinkage values at 420 days.[12]

Drying shrinkage data on concrete incorporating a high percentage of silica fume, a high dosage of a superplasticizer, and moist cured for 1 and 14 days have been published by Wolsiefer.[9] The tests were performed in accordance with ASTM C157. The drying shrinkage data are shown in Table 9. No control specimens were made for comparison purposes. The drying shrinkage values were higher for specimens cured for 1 day. In blast furnace slag/silica fume concrete a slight increase in drying shrinkage results, especially when superplasticizers are used. This increase is of little practical significance.[8]

VII. PERMEABILITY

Permeability of concrete determines its resistance to chemical attack. It is well recognized that all the cementitious hydrates and some of the aggregates from which

Table 9

DRYING SHRINKAGE OF HIGH-STRENGTH
CONDENSED SILICA FUME CONCRETE[9]

	Drying shrinkage (%)			
Curing history	28 days	3 months	5 months	1 year
1-day moist	0.052	0.060	0.068	0.073
14-day moist	0.032	0.044	0.049	0.053

concretes are made are prone to attack, not only by aggressive media, but also by water. That concrete survives aqueous environments at all is attributable to the low equilibrium solubility of the hydrated components and the low rate of mass transfer in well-compacted and cured concrete. It is generally known that the less permeable the concrete, the greater its resistance to aggressive solutions or pure water. Incorporation of supplementary cementing materials such as natural pozzolans, fly ashes, and slags in concrete considerably influences its permeability.[13] The significant decrease in permeability is attributed to the influence of pozzolans on the fine pore structure and interfacial effects. As condensed silica fume is a much more efficient pozzolanic material than natural pozzolans and fly ashes, it decreases the concrete permeability dramatically. Gjorv[14] and Markestad[15] have reported the results of investigations where water permeability tests were performed on concrete in which the cement content varied from 100 to 500 kg/m^3. Concrete incorporating 100 kg/m^3 cement and 10% condensed silica fume had its permeability decreased to about 4.0×10^{-10} m/sec from 1.6×10^{-7} m/sec. The permeability of concrete containing 100 kg/m^3 cement and 20% condensed silica fume was the same as that of concrete containing a cement content of 250 kg/m^3 without silica fume. According to Gjorv,[14] at cement factors of 400 to 500 kg/m^3, water permeability was in the range of 10^{-14} to 10^{-15} m/sec for concrete with or without silica fume. This indicates that at higher cement contents silica fume has only a marginal effect on permeability.

From the results of the above investigations, it may be concluded that even concrete with low amounts of condensed silica fume is considerably less permeable to water at ordinary pressures.

VIII. BOND PROPERTIES

A number of investigators[16,17] have studied the bond characteristics between aggregates and silica fume-based cement pastes and mortars. The bond properties of such systems were found to be superior to that of pastes and mortar without condensed silica fume. However, little or no data are available on concrete-to-concrete bond for control and silica fume concretes. Sellevold[7] has referred to a Norwegian study which concluded that condensed silica fume is an important material in providing good bond at joints or in two layer construction.

Bürge,[18] in his work on the high-strength lightweight concrete incorporating condensed silica fume studied bond strength between the embedded steel and the concrete. The bond tests were performed according to ASTM Standard C234 using concrete cubes of 120 × 120 mm. All test specimens were cured at 20°C and 95% relative humidity for 28 days. The embedded steel consisted of plain steel bars of diameter of 12 mm with an embedment length of 120 mm. Some selected test data are shown in Figures 11 and 12. At lower silica fume content the increase in bond stress is marginal; only when silica fume content exceeds 20% is there a noticeable increase in the bond

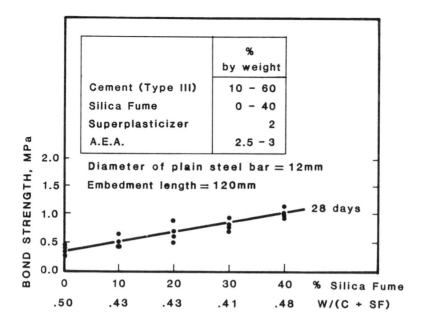

FIGURE 11.   Effect of the dosage of silica fume on the bond stress.

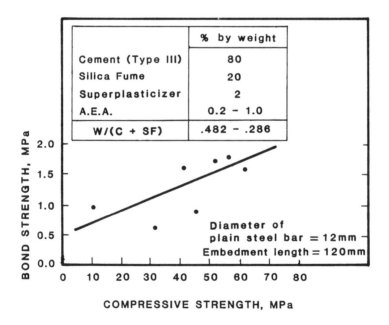

FIGURE 12.   Relation between bond stress and compressive strength of silica fume concrete.

stress value. Similar trends have been reported for normal weight concretes.[18] The compressive strength vs. bond stress shown in Figure 12 indicates that at strengths of up to about 50 MPa, no firm conclusions can be drawn on the relationship, but beyond 50 MPa the increase in bond strength follows the increase in compressive strength.

REFERENCES

1. Jahren, P., Use of Silica Fume in Concrete, ACI Spec. Publ. SP-79, Malhotra, V. M., Ed., American Concrete Institute, Detroit, 1983, 625.
2. Carette, G. G. and Malhotra, V. M., Mechanical properties, durability and drying shrinkage of Portland cement concrete incorporating silica fume, *ASTM Concrete Aggregates,* 5(1), 3, 1983.
3. Malhotra, V. M. and Carette, G. G., Silica fume concrete: properties, applications and limitations, *ACI Concrete Int., Design Constr.,* 5(5), 42, 1983.
4. Malhotra, V. M., Mechanical Properties and Freezing and Thawing Resistance of Non Air-Entrained and Air-Entrained Condensed Silica Fume Concrete Using ASTM Test C 666 Procedures A and B, Div. Rep. MRP/MSL 84-153 (OP&J), CANMET, Energy, Mines and Resources Canada, Ottawa, 1984.
5. Sellevold, E. J. and Radjy, F. F., Condensed Silica Fume (Microsilica) in Concrete: Water Demand and Strength Development, ACI Spec. Publ. SP 79, Malhotra, V. M., Ed., American Concrete Institute, Detroit, 1983, 677.
6. Skrastins, J. I. and Zoldners, N. G., Ready-Mixed Concrete Incorporating Condensed Silica Fume, ACI Spec. Publ. SP-79, Malhotra, V. M., Ed., American Concrete Institute, Detroit, 1983, 813.
7. Sellevold, E. J., Review: Microsilica in Concrete, Proj. Rep. No. 08037-EJS TJJ, Norwegian Building Research Institute, Oslo, 1984.
8. Malhotra, V. M., Carette, G. G., and Aitcin, P. C., Mechanical properties of Portland cement concrete incorporating blast-furance slag and condensed silica fume, Proc. RILEM-ACI Symp. on Technology of Concrete When Pozzolans, Slags and Chemical Admixtures are Used, Monterrey, Mexico, 1985, 395.
9. Wolsiefer, J., Ultra high-strength field placeable concrete with silica fume admixture, *ACI Concrete Int. Design Constr.,* 6(4), 25, 1984.
10. Johansen, R., Report 6: Long-Term Effects, Rep. STF 65 A 81031, Cement and Concrete Research Institute, The Norwegian Institute of Technology, Trondheim, 1983.
11. Loland, K. E. and Hustad, T., Report 2: Mechanical Properties, Rep. STF 65 A 81031, Cement and Concrete Research Institute, The Norwegian Institute of Technology, Trondheim, 1981.
12. Carette, G. G. and Malhotra, V. M., A Note on Drying Shrinkage of Condensed Silica Fume Concrete, Div. Rep. MRP/MSL 86-24 (OP&5) CANMET, Energy, Mines and Resources Canada, Ottawa, 1985.
13. Berry, E. E. and Malhotra, V. M., Fly Ash in Concrete, CANMET Spec. Publ. SP85-3, CANMET, Energy, Mines and Resources Canada, Ottawa, 1985.
14. Gjorv, O. E., Durability of Concrete Containing Condensed Silica Fume, ACI Spec. Publ. SP79, Malhotra, V. M., Ed., American Concrete Institute, Detroit, 1983, 695.
15. Markestad, A., An Investigation of Concrete in Regard to Permeability Problems and Factors Influencing the Results of Permeability Tests, Research Rep. STF 65 A 77027, STF Div. 65, The Norwegian Institute of Technology, Trondheim, 1977.
16. Regourd, M., in *Condensed Silica Fume,* Aitcin, P. C., Ed., Université de Sherbrooke, Quebec, 1983, 20.
17. Charles-Gibergues, A., Grandet, J., and Oliver, J. P., Contact zone between cement paste and aggregate, in *Bond in Concrete,* Bartos, P., Ed., Applied Science, London, 1982.
18. Bürge, T. A., High Strength Lightweight Concrete with Condensed Silica Fume, ACI Spec. Publ. SP 79, Malhotra, V. M., Ed., American Concrete Institute, Detroit, 1983, 731.

Chapter 9

DURABILITY ASPECTS OF CONDENSED SILICA FUME CONCRETE

I. INTRODUCTION

According to the ACI Committee 201 on Durability of Concrete, durability of Portland cement concrete is defined as its ability to resist weathering action, chemical attack, abrasion, or any other process of deterioration; i.e., durable concrete will retain its original form, quality, and serviceability when exposed to the environment. Engineers usually regard durability of concrete to mean its long-term satisfactory performance under severe exposure conditions whether these are natural or man-made.

In recent years this property of concrete has gained added significance and importance because of the severe degradation of concrete parking structures and bridge decks due to the combined action of deicing salts and freezing and thawing phenomenon, and failure of reinforced concrete structural elements due to the expansive chemical reaction between alkalis in cement and reactive silica in aggregates. Because of its very nature, durability of concrete cannot be easily measured or quantified in laboratory experiments and the exposure of concrete to actual service conditions takes too long to yield meaningful data for structural design purposes. In view of this, a large number of accelerated tests have been developed in various countries to provide a basis for the prediction of the long-term performance of concrete in service. However, great caution has to be exercised to interpret the results of these laboratory tests because these tests may not be universally applicable. Also, the same type of tests developed in different countries may not measure the same parameters because of the different assumptions made during the development stage. For example, the rapid freezing in water and thawing in water test as given in Procedure A of ASTM C666, though satisfactory for determining the frost resistance of concrete in hydraulic structures, may be too severe for many other applications where concrete may be exposed to frequent periods of drying. Similarly, the rapid freezing in air and thawing in water test as given in Procedure B of ASTM C666 may not be severe enough for estimating the durability of concrete exposed to the combined effect of marine environment and freezing and thawing. Furthermore, the scope of ASTM C666 states that "both procedures are intended for use in determining the effects of variations in the properties of concrete on the resistance of the concrete to the freezing and thawing cycles specified in the particular procedures. Neither procedure is intended to provide a quantitative measure of the length of service that may be expected from a specific type of concrete."

This is equally true of the various accelerated tests to determine the potential of alkali reactivity of aggregates, and the performance of concrete exposed to aggressive chemicals and sulfate attack.

It is therefore emphasized in what follows that due consideration should be given to the nature and type of tests being used by different investigators when comparing the results of their investigations.

II. FREEZING AND THAWING RESISTANCE

Extensive experience in North America has shown that for satisfactory performance of concrete under repeated cycles of freezing and thawing, the cement paste should be protected by incorporating air bubbles using an air-entraining admixture. This concept of entraining air in concrete was a chance discovery in the late 1930s. During this period it was observed that certain concrete pavements in New York State withstood well the repeated cycles of freezing and thawing. Investigations into this extraordinary behavior revealed that cements used in the manufacture of the pavement concrete had entrained air. Whether the air entrainment in the cements used was a result of the use of grinding aids in the manufacture of cement or whether it was the resut of leaked grease from the faulty bearings of the grinding mill is still open to question. Nevertheless, it provided concrete technologists with a means to produce frost-resistant concrete. Today all Canadian and U.S. specifications make the use of air entrainment mandatory for concrete exposed to frost action. A number of theories have been advanced to explain the mechanism of frost resistance of concrete by the use of air entrainment. Some of the well-known theories are the hydraulic pressure theory, the water diffusion theory, and the water vapor pressure diffusion concept.[1]

Briefly, the most important parameters concerning the entrainment of air in concrete are the air content, bubble spacing factor $\bar{L}$, and specific surface. It is generally accepted that for satisfactory freezing and thawing resistance, air-entrained concrete should have bubble spacing factor ($\bar{L}$) values of less than 200 μm (0.008 in.), and specific surface α greater than 24/mm (600/in.). Normally fresh concrete incorporating between 4 and 7% entrained air will yield the above $\bar{L}$ and α values.

The use of condensed silica fume in concrete causes changes in the microstructure and pore size distribution of the cement-silica fume binder system, and results in more impermeable matrix. This led to the speculation that perhaps concrete incorporating condensed silica fume need not incorporate entrained air for frost resistance and the published literature contains some data to this effect. One such study has been reported by Sorensen,[2] in which he concluded that nonair-entrained concrete produced with 300 kg/m^3 cement and 30 kg/m^3 of condensed silica fume exhibited outstanding frost resistance after 25 cycles of freezing and thawing as compared with a conventional concrete which failed completely. The freezing and thawing test used was based on RILEM Recommendation CDC 2: "Methods of Carrying Out and Reporting Freeze-Thaw Tests on Concrete with Deicing Chemicals". Gjorv[3] also cites somewhat similar data. The reported good performance of nonair-entrained silica-fume concrete in freezing and thawing cycling by Norwegian researchers is probably due to the type of freezing and thawing tests used and the low number of cycles to which the test specimens had been subjected.

Several studies on the freezing and thawing resistance of silica fume concrete have been reported by Carette and Malhotra[4] and Malhotra[5] using ASTM Standard C666. The method is designated as rapid because it allows for alternatively lowering the temperature of specimens from 4.4 to $-17.8°C$ and raising it from -17.8 to $4.4°C$ in not less than 2 nor more than 4 hr. The customary accepted duration of testing is 300 cycles, which can be completed in 25 to 50 days. In Procedure A of ASTM C666, both freezing and thawing occur with the specimens surrounded by water, while in Procedure B of the test the specimens freeze in air and thaw in water. Procedure B is less severe than Procedure A. The requirements for Procedure A are met by confining the specimen and surrounding water in a suitable container. The specimens used normally are prisms not less than 76 mm nor more than 137 mm in width and depth, and between 356 and 406 mm in length. The mix proportions, properties of fresh concrete, and the freezing and thawing test results for one such study are shown in Tables 1 to 4. The air-entrained concrete incorporated 0 to 30% condensed silica fume by weight as replacement for Portland cement. The W/(C + SF) was maintained at 0.40, and any loss in slump due to the use of silica fume was compensated for by the use of naphthalene-based superplasticizer. The test data revealed satisfactory performance of air-entrained concrete except for those concretes which contained 20 and 30% silica fume (Table 1,

Table 1

MIX PROPORTIONS OF CONCRETE FOR FREEZING AND THAWING
TESTS[4]

| Mix | Replacement of cement by silica fume (%) | W/(C + SF)[a] | A/(C + SF)[b] | Quantities (kg/m³) | | AEA (ml/m³)[c] | SP[d], % by wt of (C + SF) |
				Cement	Silica fume		
1	0	0.40	4.43	400	0	170	0.0
2	5	0.40	4.42	381	20	190	0.1
3	10	0.40	4.40	367	41	240	0.8
4	15	0.40	4.38	342	61	540	1.0
5	20	0.40	4.37	322	81	770	1.9
6	30	0.40	4.33	285	122	1090	2.7

[a] Water/(cement + silica fume) (by weight).
[b] Aggregate/(cement + silica fume) (by weight).
[c] Air-entraining admixture.
[d] Superplasticizer.

Table 2

PROPERTIES OF FRESH CONCRETE IN FREEZING AND
THAWING TESTS[4]

| Mix | Type | W/(C + SF)[a] | Properties of fresh concrete | | | |
			Temp. (°C)	Slump (mm)	Unit wt (kg/m³)	Air content (%)
1	Control	0.40	21	75	2340	5.1
2	5% SF + SP[b]	0.40	23	65	2330	4.4
3	10% SF + SP	0.40	22	60	2370	3.8
4	15% SF + SP	0.40	22	75	2330	4.5
5	20% SF + SP	0.40	24	180[c]	2330	4.6
6	30% SF+ SP	0.40	23	160[c]	2340	4.2

[a] Water/(cement + silica fume) (by weight).
[b] SP = superplasticizer.
[c] Increased slumps are caused by the increased dosage of the superplasticizer.

Figures 1 and 2). The excessive expansions and low value of the relative dynamic modulus shown in Figure 1 and 2 occurred in spite of the fact that the $\bar{L}$ and α values (Table 4) obtained were satisfactory. The authors of the investigation had speculated that the poor performance of the concretes in question was due to the high amount of condensed silica fume used in concrete, resulting in a very dense cement matrix that, in turn, might have adversely affected the movement of water.[4]

In another investigation, the frost resistance of nonair-entrained and air-entrained concrete incorporating various percentages of silica fume was compared.[5] The mix proportioning data and the properties of fresh concrete are given in Tables 5 and 6, and the results of the freezing and thawing tests conducted in accordance with ASTM C666 Procedures A and B are shown in Tables 7 to 9 and Figure 3. Once again, the air-entrained concrete incorporating 30% silica fume failed to meet the durability criterion (Figure 3), though in this case the poor performance of the concrete was attributed to the unsatisfactory $\bar{L}$ values. The above study led to the following conclusions:

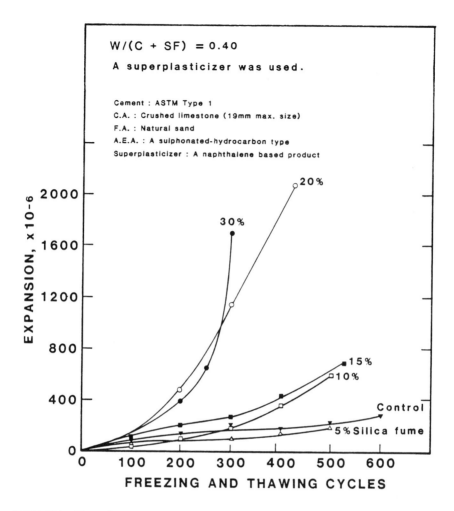

FIGURE 1.   Expansion of test prisms after freezing and thawing exposure (ASTM C666 Procedure A).[4]

Nonair-Entrained Concrete: Non air-entrained concrete, regardless of the W/(C + SF), and irrespective of the amount of condensed silica fume shows very low durability factors and excessive expansion when tested in accordance with ASTM C666 (Procedure A or B). The concrete appears to show somewhat increasing distress with increasing amounts of the fume. Therefore, the use of nonair-entrained condensed silica fume concrete is not recommended when it is to be subjected to repeated cycles of freezing and thawing.

Air-Entrained Concrete: Air-entrained concrete, regardless of the W/(C + SF) and containing up to 15% condensed silica fume as partial replacement for cement, performs satisfactorily when tested in accordance with ASTM C666 Procedures A and B. However, concrete incorporating 30% of the fume and a W/(C + SF) of 0.42, performs very poorly (durability factors less than 10) irrespective of the procedure used. This is probably due to the hardened concrete having high values of $\overline{L}$ or due to the high amount of condensed silica fume in concrete resulting in a very dense cement matrix system that, in turn, might have adversely affected the movement of water. It was difficult to entrain more than 5% air in the above type of concrete and this amount of air may or may not provide satisfactory $\overline{L}$ values in hardened concrete for durability

Table 3
SUMMARY OF TEST RESULTS AFTER FREEZING AND THAWING[4]

At the beginning of freezing and thawing cycles

Mix	Type	W/(C + SF)[a]	Weight (kg)	Length (mm)[b]	Longitudinal resonant frequency (Hz)	Pulse velocity (m/sec)
1	Control	0.40	7.402	370.6	5175	4660
2	5% SF + SP	0.40	7.335	326.9	5200	4670
3	10% SF + SP	0.40	7.350	367.8	5250	4720
4	15% SF + SP	0.40	7.327	351.0	5125	4540
5	20% SF + SP	0.40	7.285	311.4	5150	4590
6	30% SF + SP	0.40	7.224	337.6	5150	4660

At the end of freezing and thawing cycles

No. of cycles	Weight (kg)	Length (mm)	Longitudinal resonant frequency (Hz)	Pulse velocity (m/sec)	Length change (%)[b]	Relative dynamic modulus (%)[c]	Residual flexural strength (%)[c]
600	7.289	380.5	5200	4780	0.028	101	74
500	7.286	333.2	5200	4730	0.018	100	75
500	7.329	389.1	4900	4610	0.059	87	66
525	7.289	375.2	5000	4530	0.068	95	78
425	7.251	385.3	4050	3810	0.206	62	49
300	7.205	398.3	4250	3820	0.170	68	32

[a] Water/(cement + silica fume) (by weight).

[b] Gauge length = 358 mm.

[c] At the end of respective cycling, residual strength was determined in relation to reference moist-cured specimens of same age.

Table 4
AIR-VOID CHARACTERISTICS OF HARDENED CONCRETE FOR FREEZING AND THAWING TESTS[4]

Mix	Type	Air content (%)[a]	Air/ paste ratio	Specific surface (mm⁻¹)	Spacing factor $\bar{L}$ (µm)
1	Control	4.5(5.1)	0.176	22.3	221
2	5% SF + SP	4.4(4.4)	0.166	29.4	172
3	10% SF + SP	3.6(3.8)	0.135	16.3	340
4	15% SF + SP	4.3(4.5)	0.163	19.0	268
5	20% SF + SP	4.6(4.6)	0.178	17.1	285
6	30% SF + SP	4.4(4.2)	0.172	17.3	288

[a] Values in parentheses refer to air content of fresh concrete.

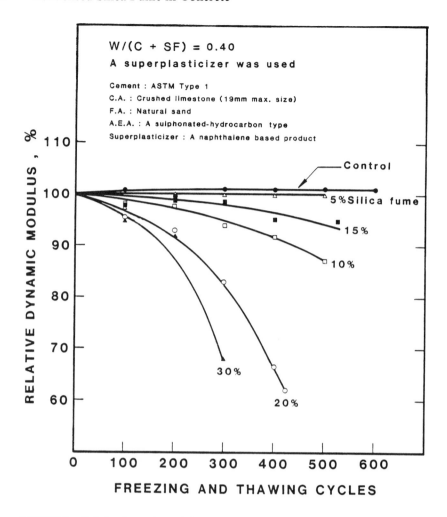

FIGURE 2.   Relative dynamic moduli of test prisms after freezing and thawing exposure (ASTM C666 Procedure A).[4]

purposes. The users are therefore asked to exercise caution when using high percentages of condensed silica fume as replacement for Portland cement in concretes with W/(C + SF) of the order of 0.40, if these concretes are to be subjected to repeated cycles of freezing and thawing.

One recent study reported results of an investigation dealing with freezing and thawing resistance of concrete in which nonair-entrained concrete was proportioned to have a W/(C + SF) ranging from 0.35 to 0.25.[6] One series of the mixes incorporated 10 and 20% silica fume. Large quantities of a superplasticizer were added to obtain slumps in the order of 120 mm (Table 10). The nonair-entrained concrete test specimens, with and without the fume, when exposed to rapid freezing and thawing tests (ASTM C666 Procedure A) started showing distress at less than 30 cycles and had developed major cracks at 50 cycles. No microcracks or scaling was observed in any instance (Figure 4). The visual observations of the width of the cracks indicated that the prisms without the fume had performed marginally better. The failure of the very low W/(C + SF) concretes in the above freezing and thawing test is probably due to the availability of freezable water in the pores of the high strength, very low permeability matrix having unsatisfactory strain capacity.

MIX PROPORTIONS AND PROPERTIES OF NONAIR-ENTRAINED FRESH
CONCRETE — FREEZING AND THAWING INVESTIGATIONS[5]

| | | | | Silica fume | | Dosage of | | | Unit wt | Air |
Mix no.	W/(C + S) (by wt)	A/C (by wt)	Cement (kg/m³)	(%)[a]	(kg/ m³)	superplasticizer (kg/m³)	Temp. (°C)	Slump (mm)	(kg/ m³)	content (%)
1	0.60	6.9	284	0	—	—	19	75	2410	1.7
2	0.60	6.9	267	5	14	—	19	85	2390	2.2
3	0.60	6.9	254	10	29	1.8	20	85	2395	1.8
4	0.60	6.9	239	15	43	2.7	20	95	2385	1.7
5	0.50	5.6	340	0	—	—	21	100	2420	1.6
6	0.50	5.6	323	5	17	0.9	21	90	2410	1.9
7	0.50	5.6	305	10	34	2.4	22	85	2400	1.9
8	0.50	5.6	289	15	51	3.5	22	85	2395	1.8
9	0.40	4.2	431	0	—	—	21	85	2420	1.8
10	0.40	4.2	415	5	22	2.8	22	75	2435	2.1
11	0.40	4.2	389	10	43	4.0	23	130[b]	2410	1.6
12	0.40	4.2	367	15	65	5.8	22	115[b]	2395	1.5
13	0.40	4.1	302	30	130	13.4	21	215	2385	1.3

[a] Percentage replacement of cement by weight.
[b] Mix no. 11, slump = 130 mm after 6 min mixing, however, was 85 mm at casting, about 15 min later; mix no. 12, slump = 115 mm after 6 min mixing, however, was 70 mm at casting, about 15 min later.

Virtanen[7] has compared the frost resistance of condensed silica-fume concrete with other concretes incorporating fly ash and blast furnace slag. Both air-entrained and nonair-entrained concretes had been used and the frost resistance of the concrete was evaluated using a battery of tests such as protective pore ratio, freezing expansion, frost-salt test at the ages of 7 and 35 days, critical saturation factor, and optical analysis of pore structure. One obvious omission was test Procedure A of ASTM C666. Test results were given ranking numbers and are shown in Table 11. According to Virtanen, if the frost resistance of concretes under investigation is considered to correlate with the total of ranking numbers in different tests, the ranking would be as shown in Table 12. The data by Virtanen[7] are generally in agreement with those reported in References 5 and 6.

III. ATTACK BY SULFATES AND OTHER CHEMICALS

The increased chemical resistance of Portland cement concrete incorporating condensed silica fume is primarily due to the condensed silica fume combining with the lime liberated during the hydration of Portland cement, thus considerably reducing the lime available for leaching. The secondary reason for the increased chemical resistance is that condensed silica fume decreases the permeability of the binder system by modifying the pore structure of mortar phase of the concrete.

In 1952, Bernhardt[8] in Norway was the first to publish limited data on the sulfate resistance of concrete exposed to 10% sodium sulfate solutions. The conclusions were that the sulfate resistance was improved when 10 to 15% of Portland cement was replaced by the condensed silica fume.

In 1971—1973 Fiskka[9,10] published data on the long-term performance of concrete test specimens exposed to ground waters containing up to 4 g/ℓ of SO_3, and pH of the

Table 6

MIX PROPORTIONS AND PROPERTIES OF AIR-ENTRAINED FRESH CONCRETE — FREEZING AND THAWING INVESTIGATIONS[5]

Mix no.	W/(C + SF) (by wt)	A/C (by wt)	Cement (kg/m³)	Silica fume (%)[a]	Silica fume kg/m³	SP[b] (kg/m³)	AEA (cm³/m³)	Temp. (°C)	Slump (mm)	Unit wt (kg/m³)	Air content (%)
						Admixture		Properties of fresh concrete			
1	0.60	7.85	243	0	—	—	137	21	75	2295	6.5
2	0.60	7.84	224	5	11	1.6	470	22	75	3315	7.0
3	0.60	7.81	221	10	23	2.6	314	21	125	2295	6.8
4	0.60	7.78	202	15	36	3.1	490	22	100	2235	7.6
5	0.40	4.28	417	0	—	—	333	21	70	2370	5.1
6	0.40	4.26	393	5	21	1.8	589	21	75	2345	5.5
7	0.40	4.25	364	10	40	3.2	1569	22	115	2280	6.9
8	0.40	4.23	340	15	61	4.0	2746	22	125	2255	7.0
9	0.42	4.17	288	30	123	11.3	7847	20	230	2295	4.8

[a] Percentage replacement of cement by weight.
[b] Superplasticizer.

Table 7
RELATIVE DYNAMIC MODULI OF ELASTICITY AND DURABILITY FACTORS AFTER VARIOUS CYCLES OF FREEZING AND THAWING: NONAIR-ENTRAINED CONCRETE[5]

Mix no.	W/(C + S) (by wt)	Air content (%)	ASTM C666, Procedure A			ASTM C666, Procedure B		
			Freezing and thawing cycles	Relative dynamic moduli (%)	Durability factor[a]	Freezing and thawing cycles	Relative dynamic moduli (%)	Durability factor[a]
1	0.60	1.7	45	71	11	44	57	8
2	0.60	2.2	45	40	6	242	7	6
3	0.60	1.8	62	7	2	99	9	3
4	0.60	1.7	62	6	1	99	9	3
5	0.50	1.6	38	N/A	N/A	94	7	2
6	0.50	1.9	55	6	1	106	10	4
7	0.50	1.9	55	10	2	106	15	5
8	0.50	1.8	38	7	<1	94	12	4
9	0.40	1.8	35	12	1	51	23	4
10	0.40	2.1	35	10	1	51	34	6
11	0.40	1.6	35	21	1	51	8	1
12	0.40	1.5	35	22	2	51	20	3
13	0.40	1.3	30	14	1	50	14	2

[a] These values have been calculated on the basis of the completion of 300 cycles of freezing and thawing.

Table 8

RELATIVE DYNAMIC MODULI OF ELASTICITY AND DURABILITY FACTORS
AFTER VARIOUS CYCLES OF FREEZING AND THAWING: AIR-ENTRAINED
CONCRETE[5]

Mix no.	W/(C + S) (by wt)	Air content (%)	ASTM C666, Procedure A			ASTM C666, Procedure B		
			Freezing and thawing cycles	Relative dynamic moduli (%)	Durability factor	Freezing and thawing cycles[a]	Relative dynamic moduli (%)	Durability factor[b]
1	0.60	6.5	320	98	100	266	97	>86
2	0.60	7.0	320	95	100	266	97	>86
3	0.60	6.8	320	96	100	266	98	>87
4	0.60	7.6	320	98	100	266	100	>89
5	0.40	5.1	309	102	100	217	100	>72
6	0.40	5.5	309	101	100	217	100	>72
7	0.40	6.9	309	99	100	217	100	>72
8	0.40	7.0	309	98	100	217	101	>72
9	0.42	4.8	75	9	2	92	24	7

[a] Tests had to be terminated at the indicated cycles because of the failure of the freezing and thawing unit.
[b] The durability factors would have exceeded the values shown if there had been no breakdown of the equipment.

Table 9

AIR-VOID ANALYSIS OF HARDENED CONCRETE — FREEZING AND THAWING INVESTIGATIONS[5]

Mix no.	W/(C + SF) (by wt)	Replacement of cement by silica fume (%)	Air content of fresh concrete (%)	Air-void parameters of hardened concrete			
				Voids in concrete (%)	Voids/mm	Specific surface (mm⁻¹)	Spacing factor (mm)
1	0.60	0	6.5	7.4	0.434	23.5	0.104
2	0.60	10	6.8	7.0	0.382	19.3	0.147
3	0.60	15	7.6	7.6	0.408	21.8	0.112
4	0.40	0	5.1	6.3	0.496	31.6	0.109
5	0.40	10	6.9	7.5	0.500	26.6	0.120
6	0.40	15	7.0	7.4	0.494	26.1	0.136
7	0.42	30	4.8	3.9	0.097	10.2	0.569

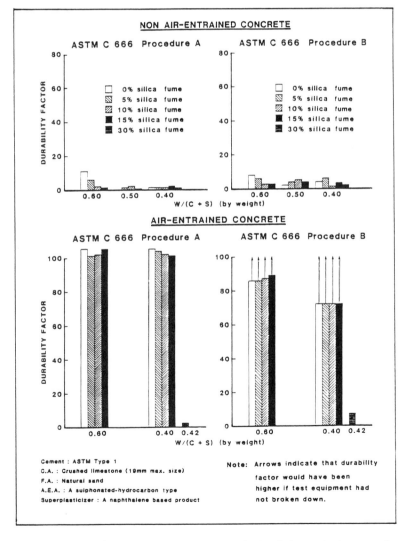

FIGURE 3.   Durability factors for nonair-entrained and air-entrained concrete.[5]

Table 10

MIX PROPORTIONS AND PROPERTIES OF FRESH AND HARDENED HIGH-STRENGTH CONCRETE[6]

Mix no.	W/(C + SF)	A/(C + SF)	Cement (kg/m³)	Silica fume %	Silica fume kg/m³	Super plasticizer (kg/m³)	Slump (mm)	Air content (%)	Density (kg/m³)	Compressive strength at 28 days (MPa)	No. of cycles completed	Durability factor
1	0.35	5.2	377	0	0	4.8	165	2.0	2460	51.5	66	6
2	0.35	5.1	342	10	38	6.2	140	1.8	2460	61.4	70	6
3	0.35	5.1	306	20	77	9.4	165	1.2	2470	68.5	70	10
7	0.30	4.3	443	0	0	7.6	215	1.9	2480	61.9	67	12
8	0.30	4.3	400	10	44	9.0	195	1.3	2475	79.3	67	3
9	0.30	4.2	356	20	89	12.5	230	1.1	2460	80.0	70	3
13	0.25	3.4	531	0	0	10.4	230	2.0	2490	65.9	89	11
14	0.25	3.4	482	10	54	11.9	230	1.3	2490	81.7	89	5
15	0.25	3.4	428	20	107	14.2	215	1.5	2465	87.1	89	8

Note: All ratios are by weight; compressive strength was determined on 150 × 300-mm cylinders; freezing and thawing test used was Procedure A of ASTM C666; cement type — ASTM Type 1, coarse aggregate — crushed limestone, fine aggregate — natural sand, superplasticizer — a naphthalene-based product; freezing and thawing test results indicate complete failure of test prisms at less than 100 cycles.

FIGURE 4. Nonair-entrained concrete prisms with and without condensed silica fume after 79 freezing and thawing cycles (ASTM C666 Procedure A). All W/C ratios = 0.25. (Top) 0% silica fume, (middle) 10% silica fume, (bottom) 20% silica fume.[6]

Table 11

SUMMARY OF FROST-RESISTANCE TESTS: NORWEGIAN INVESTIGATIONS[7]

					Ranking numbers				
Mix no.	Concrete	Air content (%)	Protective pore ratio	Freezing expansion	Frost-salt test (7 days)	Frost-salt test (35 days)	Service life	Spacing factor	Total
C-4	Cement	7.0	1	6	1	2	1	2	13
F-4	Fly ash	6.2	3	2	4	8	4	1	22
F-3	Fly ash	5.2	4	4	6	6	5	3	28
C-3	Cement	5.0	4	7	5	2	6	4	28
S-2	Silica	4.6	2	3	1	4	3	8	21
F-2	Fly ash	4.2	7	1	7	7	9	6	37
B-3	Slag	4.1	6	5	3	5	2	5	26
C-2	Cement	3.7	8	8	9	1	7	6	39
B-2	Slag	2.0	8	9	7	10	8	9	51
C-1	Cement	1.5	10	10	12	12	13	13	70
S-1	Silica	1.5	12	11	10	9	12	10	64
F-1	Fly ash	1.2	10	12	13	11	10	12	68
B-1	Slag	1.0	13	13	11	13	11	11	72

Table 12

RANKING OF DIFFERENT CONCRETES IN REGARDS TO FROST RESISTANCE: NORWEGIAN INVESTIGATIONS[7]

C-4	Cement concrete	7.0% air-entrained
S-2	Silica fume concrete	4.6% air-entrained
F-4	Fly ash concrete	6.2% air-entrained
B-3	Slag concrete	4.1% air-entrained
F-3	Fly ash concrete	5.2% air-entrained
C-3	Cement concrete	5.0% air-entrained
F-2	Fly ash concrete	4.2% air-entrained
C-2	Cement concrete	3.7% air-entrained
B-2	Slag concrete	2.0% air-entrained
S-1	Silica fume concrete	Nonair-entrained
F-1	Fly ash concrete	Nonair-entrained
C-1	Cement concrete	Nonair-entrained
B-1	Slag concrete	Nonair-entrained

waters varying from 7 to 2.5. The test specimens which had been placed in a tunnel in Oslo's alum shale region were cast from concrete mixes covering a wide range of variables. The concretes had been proportioned to have a W/C (water to cementitious materials ratio) of 0.50, except that the one incorporating 15% silica fume had a W/C of 0.62 to meet the higher water demand of condensed silica fume. The volume changes were used as a criterion for the evaluation. The measurements after 20 years of exposure indicated that the most resistant concretes were those made with either sulfate resistant or Portland cement in which 15% of the cement had been replaced by the condensed silica fume. Furthermore, the performance of the specimens made from these two types of concrete were comparable. Sellevold[11] attributes the superior per-

Table 13
MIX PROPORTIONS OF LOW W/C RATIO CONCRETES: INVESTIGATIONS OF EXPOSURE TO AGGRESSIVE CHEMICALS[12]

	Type of concrete		
	LMC (kg/m³)	SFMC (kg/m³)	LWC (kg/m³)
Portland cement	391	404	488
Fine aggregate	1013	887	876
Coarse aggregate	676	887	876
Water	137	153	161
W/C ratio	0.35	0.33	0.33

Note: LMC: latex modified concrete, SFMC: silica fume concrete, LWC: low W/C ratio concrete, LMC contains 16% solid latex, and SFMC contains 15% silica fume by weight of cement.

Table 14
PROPERTIES OF LOW W/C RATIO CONCRETES: INVESTIGATIONS OF EXPOSURE TO AGGRESSIVE CHEMICALS[12]

	Type of concrete		
	LMC	SFMC	LWC
Slump (mm)	225	250	18
Air content (%)	5.6	4.1	4.9
Av. compressive strength of cylinders before immersion in solutions (MPa)	47	74	64

formance of silica fume concrete to the refined pore structure of the matrix, lower content of calcium hydroxide, and an increased amount of aluminum incorporated in the hydrates, thus reducing the amount of alumina available for ettringite production.

The relative chemical resistance of low W/C concretes, containing either styrene-butadiene latex or silica fume, has been compared when exposed to the following solutions: 1% HCl, 1% H_2SO_4, 1% lactic acid, 5% acetic acid, 5% ammonium sulfate, and 5% sodium sulfate.[12] The criterion for failure was the time taken for 25% weight loss by fully submerged concrete specimens in the aggressive solutions. The mix proportions and properties of the concrete used are shown in Tables 13 and 14, and comparative performance of the concrete cylinders, 45 × 90 mm in size, are shown in Figures 5 to 10. The investigation revealed that except for ammonium sulfate solution (Table 15) the concrete incorporating condensed silica fume generally showed better resistance to the aggressive solutions. The rationale for relative poor performance of silica fume concrete in ammonium sulfate solution is that ammonium salts are able to decompose the calcium silicate hydrate which is the principal solid phase in hydrated

Table 15
WEIGHT LOSS TEST RESULTS OF
LOW W/C RATIO CONCRETES
AFTER EXPOSURE TO
AGGRESSIVE CHEMICALS[12]

Solution	25% weight loss at 20°C (days)		
	LMC	SFMC	LWC
1% HCl	39	63	39
5% acetic acid	75	168	48
1% lactic acid	115	182	110
1% H_2SO_4	a	a	105
5% $(NH_4)_2SO_4$	a	120	120
5% Na_2SO_4	No loss	No loss	No loss

a   20% weight loss at the end of the test (182 days).

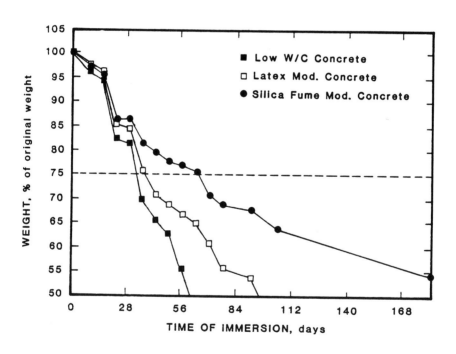

FIGURE 5.   Changes in weight of test cylinders after immersion in 1% HCl.[12]

Portland cement paste. Also, it may enhance the formation of expansion-producing ettringite. The incorporation of silica fume in concrete helps to increase the amount of calcium silicate hydrate in the system and this does not provide additional protection against attack from the ammonium salt solutions.[12] The high weight losses shown in Figures 5 to 9 are for extreme conditions of exposure, i.e., full submergence in solutions of high concentrations. In practice, such extreme exposure conditions are rarely encountered.

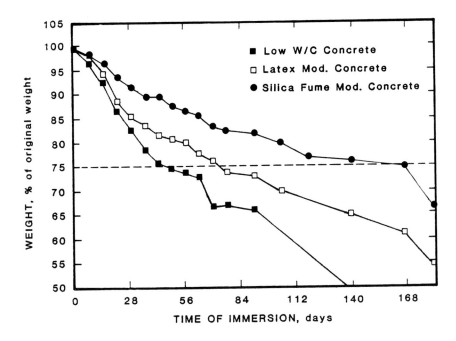

FIGURE 6.   Changes in weight of test cylinders after immersion in 5% acetic acid.[12]

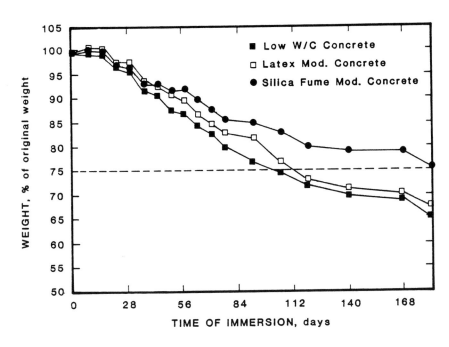

FIGURE 7.   Changes in weight of test cylinders after immersion in 1% lactic acid.[12]

The performance of concrete incorporating a silica-fume based admixture* in saturated calcium nitrate and ammonium nitrate solutions has been reported from Norway.[13] The 100-mm cubes made from concrete with and without the admixture were

* The admixture is commercially known as Corrocem and contains about 90% condensed silica fume, the
rest being a superplasticizer and Portland cement.

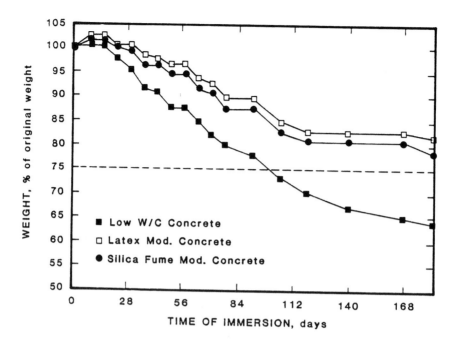

FIGURE 8.   Changes in weight of test cylinders after immersion in 1% sulfuric acid.[12]

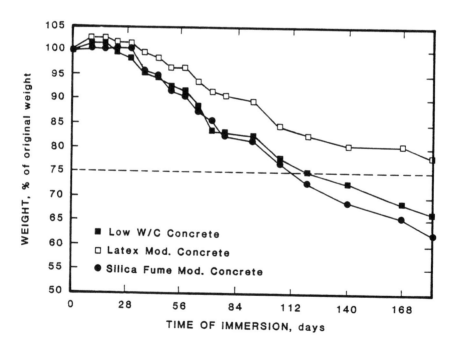

FIGURE 9.   Changes in weight of test cylinders after immersion in 5% ammonium sulfate solution.[12]

half immersed in the saturated solutions. The mixture proportioning data for the concrete used are given in Table 16. The relative loss in weight and compressive strength were used as the failure criteria. The data in Tables 17 and 18 show that whereas concrete specimens without admixture had shown weight losses of 15.1% after 108 weeks of exposure in the ammonium nitrate solutions, the specimens incorporating the

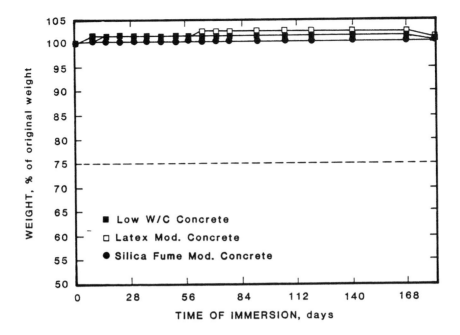

FIGURE 10. Changes in weight of test cylinders after immersion in 5% sodium sulfate solution.[12]

Table 16

MIX PROPORTIONS OF CONCRETE WITH AND WITHOUT SILICA FUME
ADMIXTURE FOR EXPOSURE TO CALCIUM NITRATE OR AMMONIUM NITRATE
SOLUTIONS[13]

Mix no.	Cement (kg/m³)	Silica fume admixture (kg/m³)	W/C	F.A. (kg/m³)	C.A. (kg/m³)	Slump (mm)	Compressive strength (MPa)			
							1-day	7-day	28-day	1064-day
1	308	20	0.50	790	1120	140	32	62	88	109
2	407	20	0.46	735	1107	120	34	68	90	109
3	310	20	0.49	790	1120	80	36	63	86	100
4	413	20	0.39	735	1107	90	43	74	97	115
5	411	—	0.47	740	1110	90	21	43	52	69
6	402	LS	0.48	740	1110	150	16	43	49	64
7	409	SNF	0.48	740	1110	120	22	42	49	65

Note: LS = Water reduced with lignosulfonate-based superplasticizer, and SNF = water reduced with naphthalene-based superplasticizer.

admixture had shown weight loss of less than 1%. The difference in strength loss for the two types of concrete was even more marked: the test specimens without the admixture had lost more than 70% of the original strength whereas those incorporating the admixture had lost less than 20% (Table 18).

It is claimed that the use of the admixtured concrete can considerably reduce the damage due to aggressive chemicals on concrete under service conditions, and thus reduce maintenance and repair costs.[13]

IV. CONTROL OF THE ALKALI-AGGREGATE REACTION IN CONCRETE

For a long time it was considered that aggregates are inert and do not take part in

Table 17
RECORDED WEIGHT LOSS (%) FOR CONCRETE WITH AND WITHOUT
THE SILICA FUME ADMIXTURE AFTER EXPOSURE TO SATURATED
CALCIUM NITRATE OR AMMONIUM NITRATE SOLUTIONS[13]

Mix no.	7-Day old concrete (weeks in solution)							28-Day old concrete (weeks in solution)						
	3	6	12	25	49	108	151	3	6	12	25	49	108	151
1	0.2	0.5	1.2	1.2	1.0	1.3	1.4	0.2	0.2	0.3	0.2	0.3	0.4	0.6
	(0.2)*	(0.4)	(0.7)	(0.8)	(0.9)	(0.8)	(0.7)	(0.2)	(0.2)	(0.4)	(0.4)	(0.5)	(0.5)	(0.6)
2	0.2	0.2	0.5	0.5	0.5	0.7	0.9	0.2	0.2	0.3	0.2	0.4	0.4	0.4
	(0.1)	(0.4)	(0.5)	(0.7)	(0.8)	(0.7)	(0.7)	(0.2)	(0.3)	(0.4)	(0.4)	(0.4)	(0.5)	(0.5)
3	0.3	0.3	0.9	1.0	1.1	1.1	1.2	0.2	0.2	0.3	0.2	0.2	0.3	0.4
	(0.2)	(0.5)	(0.6)	(0.8)	(0.8)	(1.3)	(0.6)	(0.2)	(0.3)	(0.3)	(0.2)	(0.3)	(0.3)	(0.3)
4	0.1	0.2	0.2	0.3	0.4	0.4	0.5	0.1	0.1	0.2	0.2	0.2	0.4	—
	(0.2)	(0.5)	(0.5)	(0.4)	(0.6)	(0.6)	(0.7)	(0.1)	(0.2)	(0.2)	(0.2)	(0.3)	(0.3)	—
5	1.8	3.7	6.6	12.2	14.4	19.6	—	0.7	2.0	4.5	10.7	14.4	17.0	—
	(1.0)	(2.0)	(3.3)	(5.2)	(8.9)	(16.1)	—	(0.9)	(2.0)	(2.7)	(5.0)	(8.4)	(15.1)	—
6	1.3	2.3	5.7	12.6	16.5	20.6	—	0.7	1.2	3.0	10.1	15.2	17.7	—
	(1.7)	(2.8)	(4.1)	(5.9)	(8.1)	(14.7)	—	(0.9)	(2.1)	(3.3)	(4.6)	(7.6)	(11.8)	—
7	1.2	2.6	5.9	13.4	15.9	22.9	—	0.8	1.4	2.9	10.4	16.2	18.7	—
	(1.5)	(2.3)	(3.4)	(4.8)	(7.1)	(11.9)	—	(0.7)	(1.6)	(2.5)	(4.4)	(7.5)	(13.0)	—

Note: The results with ammonium nitrate are in brackets.

Table 18
RECORDED COMPRESSIVE STRENGTH RESULTS FOR
CONCRETE WITH AND WITHOUT THE SILICA FUME
ADMIXTURE AFTER EXPOSURE TO SATURATED CALCIUM
NITRATE OR AMMONIUM NITRATE SOLUTIONS (MPa)[13]

Mix no.	7-Day old concrete (weeks in solution)					28-Day old concrete (weeks in solution)				
	0	12	21	151	163	0	12	21	151	163
1	62	85		70		88	96		94	
		(76)[a]	(70)		(57)		(80)	(83)		(67)
2	68	93		84		90	101		96	
		(86)	(84)		(65)		(82)	(90)		(75)
3	63	83		82		86	97		93	
		(76)	(77)		(51)		(82)	(83)		(69)
4	74	105		99		97	108			
		(82)	(85)		(71)		(100)	(94)		
5	43	24				53	24			
		(21)	(22)				(33)	(28)		
6	43	24				49	35			
		(21)	(18)				(29)	(30)		
7	42	24				49	36			
		(22)	(24)				(35)	(30)		

[a]   The results with ammonium nitrate are in brackets.

any chemical reactions in concrete. However, in 1940, it was reported by Stanton[14] that chemical reactions involving aggregates in concrete can result in serious damage to concrete by causing abnormal expansion and cracking. The reaction referred to by Stanton is known as the alkali-silica reaction, and involves a reaction between alkalis (Na_2O and K_2O) from the cement or from other sources, with hydroxyl and certain siliceous constituents that may be present in the aggregate. The principal products of alteration are distinctive gelatinous hydrates which expand as water is imbibed and exert pressure on the surrounding mortar. The characteristic features of the reaction are "clarified" rims on permeable particles and layered deposits of alkali-silica hydrates in voids. The classical example of an aggregate prone to alkali-silica reaction is the well-known opaline chert of California.

It has also been shown that certain carbonate rocks participate in reactions with alkalis which can produce detrimental expansion and cracking. The detrimental reaction is usually associated with argillaceous dolomitic limestone, and is designated as the expansive alkali-carbonate reaction. It has been extensively studied in Canada where it was originally recognized in Kingston, Ontario, by Swenson and Gillott.[15] According to their theory, the reaction involves destruction of dolomite crystals by the alkaline solutions and the release of entrapped active clay minerals that absorb water directly and swell. The principal reaction products include calcite and films of Ca-Mg hydroxides and silicates.

Pozzolanic admixtures, i.e., natural pozzolans and fly ash, and blast furnace granulated slags have been commonly used to control the expansion associated with the alkali-silica reaction. The limited published data available indicate that condensed silica fume, like other pozzolans, is equally effective, if not more so, in controlling such expansion. The suggested explanation for the effectiveness of the condensed silica fume is that very fine particles of the fume react rapidly with the available alkalis from the cement, leaving little or no alkalis to react with the reactive silica in the aggregate. The secondary rationale for the beneficial effect of the condensed silica fume is that, like other pozzolans, it reacts with the calcium hydroxide in the cement paste, thus lowering the pH of the pore solution.

Though a number of laboratory studies have been reported on the role of silica fume to control alkali-aggregate reactions in mortars, no published data are yet available on the use of the fume to control the previously mentioned reaction in actual concrete structures. Figure 11 shows one experimental pavement built in the province of Quebec in 1980.[16] The concrete incorporated known reactive aggregate which was limestone containing amorphous silica. This aggregate is known to cause deleterious expansion within 1 year in concrete structures. The cement content of the concrete in various parts of the pavement ranged from 140 to 405 kg/m^3 and replacement of cement by condensed silica fume ranged from 10 to 40% by weight of cement. After 3 years, a short time in terms of the alkali-silica reaction, the pavement appeared to be in good condition. The microstructure of concrete cores taken from the pavement at various intervals revealed no silica gel in the specimens cored from low cement content concrete, but some traces of the gel were found in a few locations encircling coarse aggregate particles in high cement content concrete.

V. PROTECTION OF STEEL REINFORCEMENT AGAINST CORROSION

In well-made, well-compacted and well-cured concrete, reinforcing steel should not corrode because of the high alkalinity of the pore solution of concrete (pH >13). This high alkalinity of concrete results in the formation of a protective oxide film on the surface of the reinforcement. The reinforcing steel is then said to be in a passive state

FIGURE 11.   A close-up view of an experimental concrete pavement built incorporating reactive aggregates and condensed silica fume, after 3 years in service.[16]

and the iron oxide layer or film is called a passive film. Under certain conditions of service, this protective film may be destroyed. This can occur when the pH of concrete is reduced to about 11.0 or in the presence of chlorides. This normally happens when, in porous concretes or in concretes with very little cover to reinforcement, the atmospheric carbon dioxide reacts with calcium hydroxide in the cement paste and converts the latter to calcium carbonate.

If the passive film is destroyed, the corrosion (rusting) of the reinforcing or other embedded steel begins. Rusting is an electrochemical process that requires a flow of electrical current for the chemical corrosion reactions to proceed. These electrochemical reactions take place when two dissimilar metals come in contact in the presence of oxygen and moisture. Under normal conditions iron becomes anode and the other metal becomes cathode with the oxidation taking place at the former site and the reduction of oxygen taking place at the latter site. The anodic and cathodic reactions take the following form: anodic reaction: $Fe \rightarrow Fe^{2+} + 2e$; cathodic reaction: $2H_2O + O_2 + 4e \rightarrow 4OH^-$. The overall reaction takes the form: $2Fe + 2H_2O + O_2 \rightarrow 2Fe(OH)_2$. The ferrous hydroxide, $Fe(OH)_2$ is further spontaneously oxidized to hydrated ferric oxide, $Fe_2O_3 \cdot nH_2O$.

Once the corrosion of the reinforcing steel starts, the rate of corrosion is primarily controlled by the rate of oxygen transport through the concrete and the electrical resistivity of the concrete. In the case of steel corrosion, a separate cathodic metal is not essential as separate parts of a reinforcing bar can develop anodic-cathodic galvanic cells.

Vennesland and Gjorv[17] and Gjorv[3] have reported results on the effect of up to 20% condensed silica fume by weight of cement on the rate of carbonation, electrical resistivity, and rate of oxygen transport through water saturated concrete. Some of their test data are shown in Tables 19 and 20 and Figures 12 to 15. The test methods used for the determination of electrical resistivity and oxygen transport are described elsewhere.[3,18] The general conclusions were

1.   The rate of oxygen transport through water-saturated concrete is only slightly affected.

Table 19

MIX PROPORTIONS OF SILICA FUME CONCRETE FOR STUDIES
ON CARBONATION DEPTH, ELECTRICAL RESISTIVITY, AND
OXYGEN FLUX[17]

Mix proportions (%)			Quantities (kg/m³)							
C	SF	P	Cement	Silica fume	Sand	Crushed stone	Water	Plasticizer	W/C	W/(C + SF)
100	0	0	96	—	1211	809	228	—	2.38	—
100	0	1	96	—	1205	807	200	1.0	2.09	—
100	0	2	94	—	1191	795	198	1.9	2.12	—
100	10	0	96	10	1201	801	222	—	2.32	2.11
100	10	1	96	10	1206	804	202	1.0	2.10	1.91
100	20	2	97	20	1207	804	195	1.9	2.02	1.70
250	0	0	245	—	1145	764	218	—	0.89	—
250	0	1	246	—	1150	767	196	2.5	0.81	—
250	0	2	251	—	1176	784	193	5.0	0.79	—
250	10	0	245	25	1131	754	239	—	0.97	0.89
250	10	1	248	25	1147	765	201	2.5	0.82	0.74
250	20	2	250	50	1142	761	193	5.0	0.79	0.55
400	0	0	404	—	1086	726	208	—	0.52	—
400	0	1	412	—	1109	742	198	4.1	0.49	—
400	0	2	426	—	1148	768	184	8.5	0.45	—
400	10	0	399	40	1054	703	225	—	0.56	0.51
400	10	1	414	41	1100	729	191	4.1	0.47	0.43
400	20	2	418	85	1077	719	184	8.4	0.46	0.38

Note: C = Portland cement, SF = condensed silica fume, and P = lignosulfonate-type plasticizer with 40% solids.

2. The rate of carbonation is somewhat reduced.
3. The electrical resistivity of the concrete is increased by up to 190 to 1600% for cement contents ranging from 100 to 400 kg/m³, respectively.

Thus, silica fume may be incorporated into concrete without concern for the corrosion of steel.

VI. HIGH TEMPERATURE EFFECTS ON CONDENSED SILICA FUME MORTARS AND CONCRETES

The loss of concrete strength by heating to different temperatures is due to the differential thermal expansion between cement paste and aggregates, and due to the resulting physical and chemical changes. Based upon the differential thermal analysis data for ordinary Portland cement concrete, Schneider[19] has shown that during heating the breakdown and decomposition of concrete takes place in the following manner:

Exposure temp. (°C)	Nature of breakdown
100	Loss of free water
180	Beginning of dehydration, i.e., loss of combined water
500	Decomposition of portlandite or $Ca(OH)_2$
700	Decomposition of C-S-H begins

Table 20

CARBONATION DEPTH, ELECTRICAL RESISTIVITY, AND OXYGEN FLUX DATA ON SILICA FUME CONCRETE[17]

Mix proportions[a] (%)			Depth of carbonation[b] (mm)	Change relative to mix without additions (%)	Resistivity (kohm/cm)	Oxygen flux (mol/cm²sec)10^{-13}
C	SF	P				
100	0	0	25.4	—	4.25	3.37
100	0	1	28.1	10.6	5.64	1.45
100	0	2	21.3	−16.1	5.36	2.90
100	10	0	18.2	−28.3	9.89	2.46
100	10	1	17.3	−39.9	8.90	3.24
100	20	2	11.3	−55.5	16.55	1.45
250	0	0	6.8	—	6.57	5.05
250	0	1	4.9	−27.9	6.64	4.51
250	0	2	4.8	−29.4	6.57	4.33
250	10	0	6.2	−8.8	17.58	4.33
250	10	1	3.9	−42.6	20.51	3.32
250	20	2	3.4	−50.0	47.52	0.67
400	0	0	3.7	—	6.89	3.78
400	0	1	2.7	−27.0	7.55	3.96
400	0	2	2.2	−40.5	8.89	5.44
400	10	0	3.8	2.7	28.86	5.75
400	10	1	2.1	−43.2	49.34	5.18
400	20	2	1.2	−67.6	127.24	4.64

[a] C = Portland cement, SF = condensed silica fume, and P = lignosulfonate-type plasticizer with 40% solids.

[b] The specimens were 1100 × 1100 × 100 mm prisms. After demolding, the prisms were kept humid for 7 days and then stored in a room with 60% relative humidity and temperature until 14.5 months.

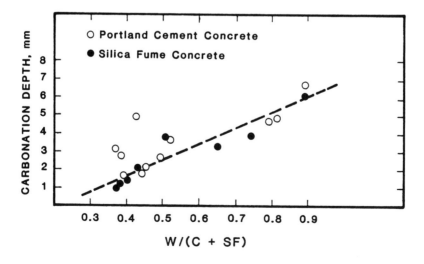

FIGURE 12.　Effect of W/(C + SF) on the carbonation depth of control and silica fume concrete.[17]

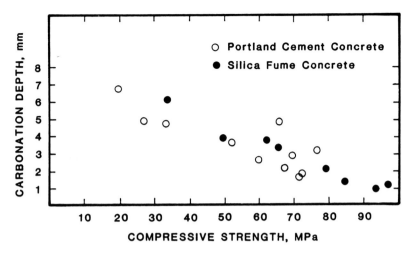

FIGURE 13. Relation between carbonation depth and compressive strength of control and silica fume concrete.[17]

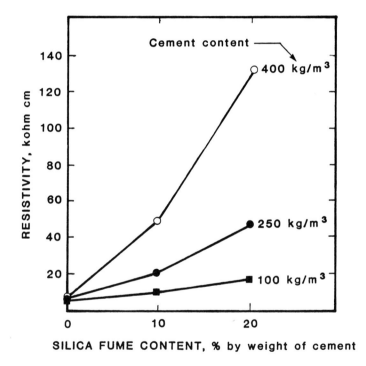

FIGURE 14. Effect of the dosage of silica fume on the resistivity of concrete.[17]

At a temperature of 700 to 800°C, calcareous aggregates if used in concrete are also decomposed. Invariably, cement paste in concrete is carbonated to some extent and this also decomposes in this temperature range.

Williamson and Rasheed[20] have used Schneider's analysis of breakdown of concrete during heating to explain the loss of strength of Portland cement mortars incorporating condensed silica fume. The data are given in Tables 21 and 22. It is seen that low-strength mortars incorporating condensed silica fume are less susceptible to strength loss after high temperature exposure than mortars without condensed silica fume. On

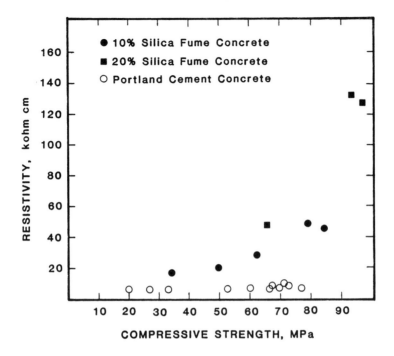

FIGURE 15.   Relation between resistivity and compressive strength of control and silica fume concrete.[17]

Table 21

MIX PROPORTIONS OF MORTARS INCORPORATING SILICA FUME
FOR HIGH TEMPERATURE EXPOSURE[20]

Mix no.	Cement (g)	Water (m*l*)	Silica fume (g)	Fine aggregate (g)	Plasticizer (m*l*)	Superplasticizer (g)
LOW-0	100	71	0	392	0.52	0.00
LOW-8	100	88	8	446	0.70	0.00
LOW-16	100	105	16	513	1.73	0.00
MED-0	100	46	0	201	0.51	0.00
MED-8	100	57	8	232	1.10	0.00
MED-16	100	67	16	273	1.73	0.00
HIGH-0	100	26	0	70	1.00	0.32
HIGH-8	100	29	8	84	1.80	0.00
HIGH-16	100	34	16	97	2.36	0.00

the other hand, high-strength mortar with condensed silica fume seems to be more susceptible to high temperatures than high-strength mortars without condensed silica fume. The former retains only 65% of its strength at room temperature after heating to 320°C in comparison with regular high-strength mortar which retains 95% of its strength at room temperature (Figures 16 to 18).

Williamson and Rasheed[20] have theorized that condensed silica fume toughens the portlandite (calcium hydroxide) in high-strength pastes, but after heating to 320°C the condensed silica fume particles which have been converted to C-S-H lose their ability to toughen the portlandite because of the partial decomposition of C-S-H at 300°C.

Data reported by Williamson and Rasheed have been corroborated by limited CAN-

Table 22
RELATIVE STRENGTH AND MODULUS OF ELASTICITY OF SILICA FUME MORTARS AFTER EXPOSURE TO DIFFERENT TEMPERATURES[20]

Relative strength of mortars

Temp.	LOW-0	LOW-8	LOW-16	MED-0	MED-8	MED-16	HIGH-0	HIGH-8	HIGH-16
20°C	Av. strength (MPa)								
	40.7	37.2	33.1	75.9	75.9	60.7	112.4	99.3	120.7
	Av. modulus of elasticity E_o (x 10^4 MPa)								
	2.97	2.65	2.51	3.18	2.87	2.35	7.3	3.23	3.25
320°C	Strength in % of value at 20°C								
	96.1	108.0	112.0	90.6	86.2	95.2	94.7	89.1	65.0
	Modulus of elasticity in % of E_o								
	66.8	63.6	49.7	83.9	65.8	44.9	57.0	58.0	37.4
520°C	Strength in % of value at 20°C								
	72.7	78.0	80.7	63.0	58.1	60.0	75.0	61.0	42.3
	Modulus of elasticity in % of E_o								
	25.0	21.2	24.1	32.5	31.0	22.3	20.0	22.1	18.0
700°C	Strength in % of value at 20°C								
	45.5	40.0	42.5	42.3	35.0	40.0	48.0	37.02	27.3
	Modulus of elasticity in % E_o								
	8.0	5.0	6.9	12.0	9.4	11.3	9.0	8.4	6.0

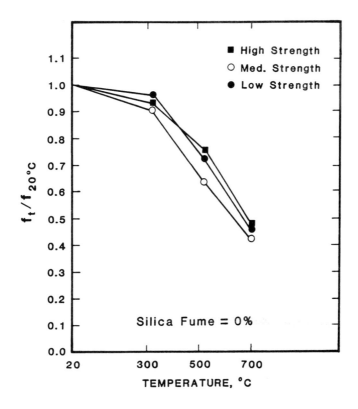

FIGURE 16. Ratio of $f_t/f_{20°c}$ vs. temperature of exposure for plain mortar test specimens.[20]

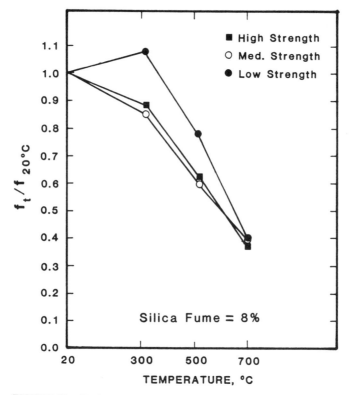

FIGURE 17.   Ratio of $f_t/f_{20°C}$ vs. temperature of exposure for mortar test specimens incorporating 8% silica fume.[20]

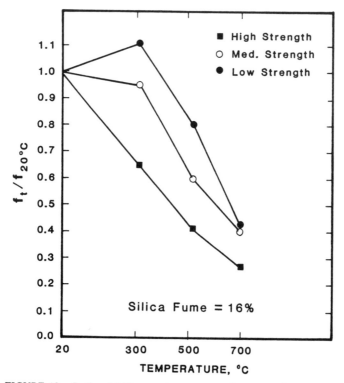

FIGURE 18.   Ratio of $f_t/f_{20°C}$ vs. temperature of exposure for mortar test specimens incorporating 16% silica fume.[20]

Table 23

CHANGES IN WEIGHT AND PULSE VELOCITY OF TEST
PRISMS SUBJECTED TO VARIOUS TEMPERATURES FOR A
SUSTAINED PERIOD OF 60 DAYS[21]

Mix no.	W/(C + SF) (by wt)	Air content (%)	Silica fume content (%)	Oven temp. (°C)	Relative change in test prisms[a] after heat expsoure (%)	
					Weight	Pulse velocity
1	0.60	1.7	0	75	−3.2	−11.5
				150	−4.0	−19.6
				300	−4.4	−36.2
2	0.60	1.7	15	75	−4.6	−15.2
				150	−5.4	−28.6
				300	−5.9	−46.5
3	0.40	1.8	0	75	−3.4	−13.8
				150	−4.1	−21.0
				300	−4.5	−38.9
4	0.40	1.5	15	75	−4.1	−14.6
				150	−5.3	−27.9
				300	−5.5	−44.8
5	0.60	6.5	0	75	−2.5	−10.3
				150	−3.0	−18.8
				300	−2.9	−38.2
6	0.60	7.6	15	75	−3.5	−17.2
				150	−3.6	−24.1
				300	−4.1	−44.3
7	0.40	5.1	0	75	−3.5	−10.1
				150	−3.5	−16.5
				300	−4.4	−38.7
8	0.40	7.0	15	75	−3.8	− 8.6
				150	−4.8	−22.0
				300	−5.1	−41.7

Note: All measurements of weights and pulse velocity were made immediately before and after exposure.

[a] Test specimens, 88 × 100 × 400 mm in size, were moist cured for 14 days followed by lab-drying for 7 days and subsequent heat exposure for 60 days.

MET data on concrete prisms exposed to sustained temperatures ranging from 75 to 300°C for a period of 60 days.[21] Ultrasonic pulse velocity and weight loss determinations on test prisms before and after sustained heat exposure show that concrete incorporating 15% condensed silica fume by weight of cement performs poorly as compared with control prisms without condensed silica fume (Table 23). This was so for both nonair-entrained and air-entrained concrete.

Sellevold[11] has summarized the results of limited Danish and Norwegian studies on exposure of condensed silica-fume concrete to high temperature and fire, and also referred to one investigation by Pedersen. In this study several small cylinders of very low W/(C + SF) (<0.20) and containing 20% silica fume suddenly disintegrated at 300°C. This type of behavior is not entirely unusual for ultrahigh strength and impermeable concrete because under slow heating, high vapor pressure can build up inside a specimen. Sellevold[11] cites another case where four large concrete elements, one incorporating silica fume, were tested for fire resistance in accordance with ISO standards. The 28-day compressive strength ranged from 32 to 35 MPa and the elements were 3 months old when tested. In this test all elements met the temperature exposure requirements for the unexposed face; however, higher spalling was noticed on the exposed face for the element incorporating silica fume.

REFERENCES

1. Dolch, W. L., Air-Entraining Admixtures, Concrete Admixtures Handbook: Properties, Science, and Technology, Ramachandran, V. S., Ed., Noyes Publications, N.J., 1984, 269.
2. Sorensen, E. V., Freeezing and Thawing Resistance of Condensed Silica Fume, (Microsilica) Concrete Exposed to Deicing Salts, ACI Spec. Publ. SP79, Malhotra, V. M., Ed., American Concrete Institute, Detroit, 1983, 709.
3. Gjorv, O. E., Durability of Concrete Containing Condensed Silica Fume, ACI Spec. Publ. SP79, Malhotra, V. M., Ed., American Concrete Institute Detroit, 1983, 695.
4. Carette, G. G. and Malhotra, V. M., Mechanical properties, durability and drying shrinkage of Portland cement concrete incorporating silica fume, *ASTM Cement Concrete Aggregates*, 5(1), 3, 1983.
5. Malhotra, V. M., Mechanical Properties and Freezing and Thawing Resistance of Non-Air-Entrained and Air-Entrained Condensed Silica Fume Concrete Using ASTM Test C 666 Procedures A and B, CANMET Div. Rep. MRP/MSL 84-153 (OP & J), CANMET, Energy Mines and Resources Canada, Ottawa, 1984.
6. Malhotra, V. M., Painter, K. E., and Bilodeau, A., Freezing and Thawing Resistance of High-Strength Concrete With and Without Condensed Silica Fume, CANMET Div. Rep. MRP/MSL 86-23(J), CANMET, Energy, Mines and Resources Canada, Ottawa, February 1986.
7. Virtanen, J., Freeze-Thaw Resistance of Concrete Containing Blast-Furnace Slag, Fly Ash or Condensed Silica Fume, ACI Spec. Publ. SP79, Malhotra, V. M., Ed., American Concrete Institute, Detroit, 1983, 923.
8. Bernhardt, C. I., SiO_2 — Stov Som Cementtilsetning, *Betongen dag*, 17(2), 1952 (in Norwegian).
9. Fiskaa, O., Hansen, H., and Moum, J., Betong i Aluskifer, Publ. No. 86, Norsk Geoteknisk Institute, Oslo, 1971 (in Norwegian).
10. Fiskaa, O. M., Betong i Alunskifer, Publ. No. 101, Norsk Geoteknisk Institute, Oslo, 1973 (in Norwegian).
11. Sellevold, E. J., Review: Microsilica in Concrete, Proj. Rep. 08037 — EJS/TJJ, Norwegian Building Institute, Oslo, 1984.
12. Mehta, P. K., Durability of Low Water-Cement Ratio Concretes Containing Latex or Silica Fume as Admixtures, Proc. RILEM-ACI Symp. on Technology of Concrete When Pozzolans, Slags, and Chemical Admixtures Are Used, Monterrey, Mexico, 1985, 325.
13. Carlsson, M., Hope, R., and Pedersen, J., Practical Benefits From Use of Silica Fume in Concrete, Departmental Report A/S Scancem, Slemmestad, Norway, 1985.
14. Stanton, T. E., Expansion of concrete through reaction between cement and aggregates, *Proc. ASCE*, 66, 1781, 1981.
15. Swenson, E. G. and Gillott, J. E., Alkali-Carbonate Rock Reaction, Highway Research Record, No. 45, Highway Research Board, Washington, D.C., 1964, 21.
16. Aitcin, P. C. and Regourd, M., The Use of Condensed Silica Fume to Control Alkali-Silica Reaction — A Field Case Study (available from University of Sherbrooke, Sherbrooke, Quebec, Canada).
17. Vennesland, O. ad Gjorv, O. E., Silica Concrete — Protection Against Corrosion of Embedded Steel ACI Spec. Publ. SP79, Vol. 2, Malhotra, V. M., Ed., American Concrete Institute, Detroit, 1983, 719.
18. Gjorv, O. E., Oystein, V., and El-Busaidy, A. H., Diffusion of Dissolved Oxygen Through Concrete, paper no. 17, NACE Corrosion 76, Houston, 1976.
19. Schneider, U., Verhalten Von Beton Bei Hohen Temperturen; *Dtsch. Ausschuss Stahlb.*, 337, 1982.
20. Williamson, R. B. and Rasheed, I. A., The Use of High-Strength Concrete and Mortars in Combustion Environment (paper obtainable from Department of Civil Engineering, University of California, Berkeley, 1984).
21. Painter, K. E., A Note on the Pulse Velocity of Silica Fume Concrete Prisms after Exposure to Elevated Temperatures, CANMET Intern. Rep. MRP/MSL 86-25(R), Energy, Mines and Resources Canada, Ottawa, February 1986.

Chapter 10

APPLICATIONS

I. INTRODUCTION

There are a number of applications in which condensed silica fume can be incorporated to advantage in concrete. These will be discussed in this chapter.

II. LOWER CEMENT CONTENT

Norway is one of the largest producers of condensed silica fume in the world, and ready-mixed concrete producers and precast concrete manufacturers in that country have been using silica fume as replacement for cement on strictly economic grounds. As mentioned earlier, the so-called efficiency factors, when silica fume is used in this manner, range from 2 to 5, depending upon the strength level of concrete and its cement factor. The use of condensed silica fume in this manner is probably due to the following factors: large quantities of the fume are available within easy reach of ready-mixed concrete producers, cement is relatively more expensive, and other pozzolanic materials are not easily available. In Quebec, Canada, condensed silica fume is also being used in the same manner as in Norway, though the supply is less than 20,000 t/ year. Figure 1 shows a set-up describing the storage and handling of silica fume at a concrete ready-mixed plant in Quebec. Some data on the relative amounts of Portland cement and condensed silica fume used for a given compressive strength level as used in a ready-mixed plant are shown in Figures 2 to 4.[1]

The experience gained in Quebec parallels that in Norway; given the economic haulage distance and the lack of other supplementary cementing materials, condensed silica fume can be used with advantage in a ready-mixed concrete operation. The data shown in Figures 2 to 4 indicate that the use of up to 12% condensed silica fume is advantageous for maintaining required strength levels and at the same time saving substantial quantities of cement. The current trend in Quebec is to replace less than 10% cement by silica fume.

III. EARLY-AGE STRENGTH OF CONCRETE MADE WITH FLY ASH OR SLAG

When low-calcium fly ashes or blast furnace slags are used as partial replacements for cement in concrete, the compressive strength of such concretes is generally lower at early ages, especially at 3 or 7 days, though some exceptions have been reported.[2] The problem can be overcome by rational mix proportioning methods which involve the incorporation of additional amounts of fly ash or slag, over and above that which has already been added as a direct replacement of cement. This approach may or may not work and will depend upon the type of supplementary cementing material being used. The preferred approach should be to use small quantities of condensed silica fume to overcome the above problem. This approach was first reported by Mehta and Gjorv.[3] In addition to performing strength tests on concrete incorporating fly ash and condensed silica fume, they also performed studies on neat cement pastes containing the same proportions of pozzolans and water. The dried paste specimens were subjected to free lime content determination by ASTM C114, and pore size distribution analysis by the mercury penetration technique. The mix proportions used and compressive

FIGURE 1. Storage and handling facilities for condensed silica fume at a ready-mixed concrete batching plant in Quebec, Canada.[1]

Potential Compressive Strength MPa	Control Mix	Field Mix		Efficiency Factor
	Cement kg/m³	Cement kg/m³	Silica Fume kg/m³	
20	252	187	22.4	2.9
25	295	207	24.8	3.5
30	335	227	25.0	4.3
35	377	246	24.6	5.3

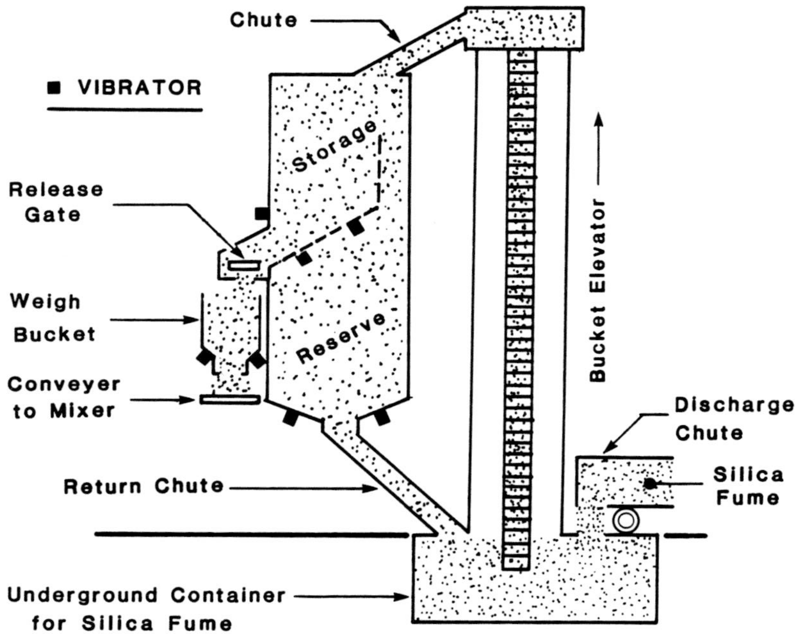

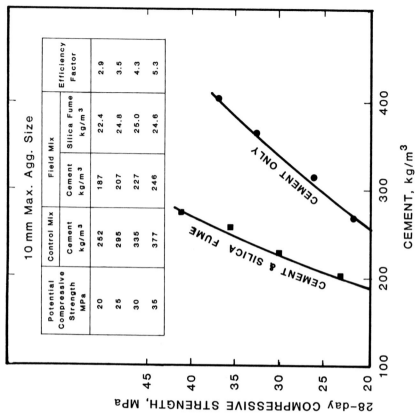

FIGURE 2. Relation between compressive strength and cement and condensed silica fume content for concrete with 10-mm maximum size aggregate.[1]

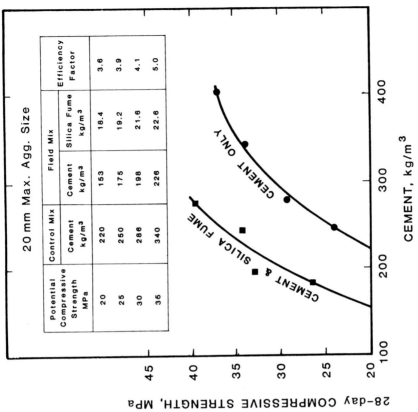

FIGURE 4. Relation between compressive strength and cement and condensed silica fume content for concrete with 20-mm maximum size aggregate.[1]

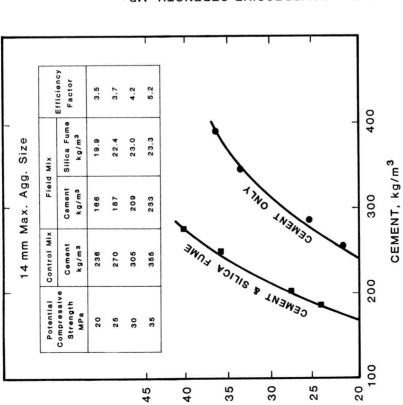

FIGURE 3. Relation between compressive strength and cement and condensed silica fume content for concrete with 14-mm maximum size aggregate.[1]

Table 1

MIX PROPORTIONS AND SLUMP OF CONCRETE
CONTAINING FLY ASH AND CONDENSED SILICA FUME[3]

	Control concrete	Concrete containing condensed silica fume	Concrete containing fly ash	Concrete containing both pozzolans
Cement content (kg/m³)	300	210	210	210
Condensed silica fume (kg/m³)	0	60	0	30
Fly ash (kg/m³)	0	0	60	30
Fine aggregate (kg/m³)	1268	683	975	683
Coarse aggregate (kg/m³)	683	1268	975	1268
Water (kg/m³)	222	202	208	205
Slump (mm)	170	25	130	50

Table 2

COMPRESSIVE STRENGTH OF CONCRETE CONTAINING FLY ASH AND
CONDENSED SILICA FUME[3]

	3 Days		7 Days		28 Days		90 Days	
	MPa	Difference from the control (%)	MPa	Difference from the control (%)	MPa	Difference from the control (%)	MPa	Difference from the control (%)
Control concrete	14.7	—	20.0	—	26.6	—	28.6	—
Concrete containing fly ash	12.1	−18	17.8	−11	23.2	−12	27.6	−3
Concrete containing condensed silica fume	15.2	+3	24.4	+22	47.0	+77	56.2	+97
Concrete containing both pozzolans	14.2	−3	20.2	+1	37.9	+42	43.3	+51

strength of concrete obtained by Mehta and Gjorv are shown in Tables 1 and 2. When using only fly ash, the strengths at 3, 7, 28, and 90 days are consistently lower than that of the reference concrete, from −18 to −3%. By the addition of silica fume to the fly ash concrete, these values change from the negative to positive values between 3 and 7 days. A net strength gain of 51% occurs at 90 days.

The free lime data in Table 3 are generally consistent with the concrete strength data in Table 2. The free lime values confirm that little or no pozzolanic activity occurs in the Portland cement/fly ash pastes until 28 days; on the contrary, considerably lower values of free lime in the pastes containing the condensed silica fume or in those with both fly ash and condensed silica fume are indicative of high pozzolanic activity even at 7 and 28 days.

The pore size distribution of the pastes at various ages can also contribute to the difference in strength characteristics of the concrete under investigation. It is found that the conversion of large pores to finer pores (<0.01 μm) as a result of pozzolanic reaction contributes to the strength enhancement of cement paste and concrete incorporating pozzolanic materials. The change in the interfacial zone between the cement paste and the aggregate is another important factor.

Table 3
FREE LIME CONTENT OF CEMENT PASTES CONTAINING CONDENSED SILICA FUME[3]

Type of paste	Percentage at		
	7 days	28 days	90 days
Control Portland cement	15.0	16.9	18.9
Portland cement containing fly ash	11.8	13.2	13.4
Portland cement containing condensed silica fume	10.7	7.9	4.2
Portland cement containing both pozzolans	11.0	10.2	9.5

Smaller quantities of condensed silica fume have also been used in both fly ash and granulated blast furnace slag concrete mixes to increase the strength of concrete at early ages.[4,5] In mixes containing fly ash, the percentage of condensed silica fume ranged from 0 to 20% by weight of (C + FA) and the W/(C + FA) of the mixes ranged from 0.40 to 0.80. The data on mix proportioning, properties of fresh concrete, and strengths are shown in Tables 4 to 6 and selected plots of the test results are given in Figures 5 to 7.

These investigations showed that the low early-age strength of Portland cement concrete incorporating fly ash can be increased by the use of condensed silica fume. The gain in strength, in general, is directly proportional to the percentage of the fume used. Specifically, at the ages of 7 days and beyond, the loss in compressive strength of the concrete due to the incorporation of fly ash can be fully compensated and even exceeded by the addition of the fume. At 7 days, the percentage of silica fume ranges from 10% for concrete with W/(C + FA) of 0.40, 0.50, and 0.60, to 15 and 20% for concretes with W/(C + FA) of 0.70 and 0.80. At 28 days, the addition is of the order of 5%, regardless of the W/(C + FA) of concrete. As for flexural strength, the above pattern of 7-day strength development is reached at 14 days.

In the investigation of the effect of silica fume on blast furnace slag (BFS) concrete, the additions of the fume ranged from 0 to 20% by weight of (cement + blast furnace slag), and the W/(C + BFS) were 0.40, 0.50, and 0.65 (Tables 7 to 9).

In general, conclusions similar to those reached for Portland cement/fly ash/condensed silica fume concrete apply for blast furnace slag concretes. Specifically, it was found that at 3 days the increase in strength was marginal, especially for concretes at high W/(C + BFS). However, at the age of 14 days and beyond, the loss in compressive strength of concrete containing blast furnace slag can be fully compensated for with a given percentage of condensed silica fume regardless of the W/(C + BFS) (Figures 8 to 10). It was also reported that increase in the drying shrinkage, resulting from the incorporation of condensed silica fume in concrete containing blast furnace slag, is marginal and of little practical consequence. This is true, of course, only when the loss in slump due to the incorporation of condensed silica fume is compensated for by the use of a superplasticizer. The time of curing required for the blast furnace slag/silica fume concrete to reach the compressive strength of reference concrete is shown in Figure 11. The results show that longer curing times are needed at low levels of silica fume. The curing times also show an increase as the W/(C + BFS) is increased. This is possibly due to the fact that at a higher W/S, the silicate hydrates formed by the reaction of silica fume are less effective in bridging the network. However, when the silica fume

Table 4
MIXTURE PROPORTIONS OF CONCRETE CONTAINING FLY ASH, CONDENSED SILICA FUME, AND A SUPERPLASTICIZER[a]

Mix no.	Type of mixture[a]	$W/(C+F)$[b]	Relative proportions of cement and fly ash (% by wt)		Condensed silica fume (% by wt of cement plus fly ash)	Batch quantities (kg/m³)					AEA (ml/m³)	Superplasticizer (% by wt of cement plus fly ash)
			Cement	Fly ash		Cement	Fly ash	Condensed silica fume	FA	CA		
1	Reference		100	0	0	382	0	0	712	1162	170	0
2	Control		70	30	0	263	113	0	691	1127	600	0
3	5% silica fume	0.40	70	30	5	263	113	18	683	1115	480	0.33
4	10%		70	30	10	265	113	39	675	1102	540	0.96
5	15%		70	30	15	263	113	56	664	1084	600	1.09
6	20%		70	30	20	262	113	75	653	1066	710	1.50
7	Reference		100	0	0	299	0	0	759	1138	80	0
8	Control		70	30	0	211	91	0	755	1131	450	0
9	5% silica fume	0.50	70	30	5	209	90	15	747	1114	430	0.30
10	10%		70	30	10	209	90	30	736	1104	440	1.00
11	15%		70	30	15	206	88	45	711	1098	640	1.49
12	20%		70	30	20	208	90	60	718	1076	410	1.47
13	Reference		100	0	0	243	0	0	811	1120	90	0
14	Control		70	30	0	171	74	0	810	1118	300	0
15	5% silica fume	0.60	70	30	5	172	74	13	806	1114	300	0.43
16	10%		70	30	10	171	74	24	799	1104	360	0.97
17	15%		70	30	15	172	74	37	797	1100	420	1.86
18	20%		70	30	20	172	74	49	788	1090	510	1.84
19	Reference		100	0	0	211	0	0	857	1090	90	0
20	Control		70	30	0	150	64	0	858	1092	320	0
21	5% silica fume	0.70	70	30	5	149	63	11	851	1083	260	0.47
22	10%		70	30	10	148	63	21	841	1071	290	0.87
23	15%		70	30	15	149	63	32	837	1064	350	1.42
24	20%		70	30	20	149	63	43	832	1058	390	1.79
25	Reference		100	0	0	185	0	0	903	1060	70	0
26	Control		70	30	0	130	56	0	902	1058	230	0
27	5% silica fume	0.80	70	30	5	131	56	10	905	1062	240	0.57
28	10%		70	30	10	131	56	18	898	1054	270	0.89
29	15%		70	30	15	130	56	28	883	1036	290	1.31
30	20%		70	30	20	130	56	37	878	1031	340	1.60

[a] Reference: 100% normal Portland cement, control: 70% normal Portland cement plus 30% fly ash, silica fume: 70% normal Portland cement plus 30% fly ash plus additions of condensed silica fume.

[b] $W/(C + FA)$ by weight.

Table 5

PROPERTIES OF FRESH CONCRETE CONTAINING FLY ASH, CONDENSED SILICA FUME, AND A SUPERPLASTICIZER[4]

Mix series, no.	Type of mixture[a]	W/(C + F)[b]	Properties of fresh concrete			
			Temp (°C)	Slump (mm)	Unit wt (kg/m³)	Air content (%)
1	Reference		22	40	2409	4.9
2	Control		22	70	2345	5.6
3	5% silica fume	0.40	21	65	2345	4.9
4	10%		22	55	2345	5.2
5	15%		21	55	2333	5.0
6	20%		21	75	2320	5.0
7	Reference		21	85	2345	6.0
8	Control		21	90	2339	6.3
9	5% silica fume	0.50	22	90	2320	6.6
10	10%		23	85	2320	6.5
11	15%		22	90	2275	6.8
12	20%		23	90	2300	6.2
13	Reference		21	70	2320	6.3
14	Control		21	90	2320	5.8
15	5% silica fume	0.60	20	95	2326	6.1
16	10%		24	70	2320	6.4
17	15%		23	70	2326	5.7
18	20%		22	70	2320	5.6
19	Reference		24	65	2307	6.4
20	Control		23	75	2313	6.2
21	5% silica fume	0.70	21	85	2307	6.5
22	10%		21	90	2294	6.5
23	15%		21	90	2294	6.5
24	20%		20	90	2294	6.0
25	Reference		21	40	2294	6.3
26	Control		22	95	2294	6.1
27	5% silica fume	0.80	22	90	2313	6.5
28	10%		21	75	2307	6.3
29	15%		21	95	2281	6.6
30	20%		21	90	2281	6.4

[a] Reference: 100% normal Portland cement, control: 70% normal Portland cement plus 30% fly ash, silica fume: 70% normal Portland cement plus 30% fly ash plus additions of condensed silica fume.

[b] W/(C + FA) by weight.

content reaches over 15% the curing times become equal at all W/S (water/solid ratios).

IV. REDUCTION OF CHLORIDE INGRESS

Limited tests conducted in the U.S. have shown that concrete incorporating condensed silica fume is considerably less permeable to chloride ion ingress. The tests were performed using the rapid chloride test method developed by the Federal Highway Administration, Washington, D.C. This method consists of monitoring the amount of electrical current passed through a test area on a concrete slab, when a potential difference of 80 V(dc) is maintained across the specimen for a period of 6 hr.[6] Chloride ions are forced to migrate out of a sodium chloride solution subjected to a negative

Table 6

COMPRESSIVE AND FLEXURAL STRENGTHS OF CONCRETE
CONTAINING FLY ASH, CONDENSED SILICA FUME, AND A
SUPERPLASTICIZER[4]

Mix no.	Type of mixture[a]	W/(C + F)[b]	Compressive strength of 152 × 305-mm cylinders (MPa)				Flexural strength of 76 × 102 × 406-mm prisms (MPa)	
			1-day	3-day	7-day	28-day	7-day	14-day
1	Reference		21.8	30.8	33.5	40.1	5.8	6.4
2	Control		11.0	18.1	24.2	33.7	4.2	5.0
3	5% silica fume	0.40	11.2	22.2	28.5	40.4	5.0	5.4
4	10%		15.6	25.7	35.3	46.8	5.4	6.8
5	15%		16.2	27.4	39.1	49.0	5.3	6.6
6	20%		16.4	28.8	41.0	53.0	6.1	6.9
7	Reference		10.7	20.1	25.5	32.0	4.6	4.9
8	Control		5.8	12.6	17.7	26.3	3.6	4.1
9	5% silica fume	0.50	5.9	13.7	19.7	30.6	3.8	4.3
10	10%		9.5	17.8	27.1	37.6	4.3	5.0
11	15%		9.3	17.7	29.1	41.5	4.0	4.9
12	20%		10.9	20.1	31.7	46.5	4.6	6.2
13	Reference		6.3	13.0	18.6	23.5	3.9	4.3
14	Control		3.4	8.3	12.4	18.5	3.0	3.7
15	5% silica fume	0.60	3.7	9.5	14.8	27.0	3.3	4.4
16	10%		6.0	12.6	18.4	31.0	3.5	4.2
17	15%		6.9	14.4	22.5	34.3	3.6	4.3
18	20%		7.0	13.6	24.4	36.6	3.8	4.9
19	Reference		5.3	10.7	14.6	19.3	3.1	3.5
20	Control		2.6	6.1	8.5	13.3	2.1	2.8
21	5% silica fume	0.70	2.3	6.5	10.6	20.4	2.2	3.1
22	10%		2.8	7.5	12.8	24.3	2.5	3.6
23	15%		3.4	8.1	14.5	27.5	2.5	3.7
24	20%		3.8	9.5	18.7	32.4	3.1	4.7
25	Reference		3.5	7.8	11.5	15.7	2.8	3.4
26	Control		1.5	4.1	6.3	10.1	1.6	2.3
27	5% silica fume	0.80	1.8	5.1	8.4	16.6	2.1	2.8
28	10%		2.1	5.4	9.3	19.6	2.0	2.9
29	15%		2.4	5.9	10.7	21.7	2.3	3.2
30	20%		2.7	6.3	12.6	22.8	2.4	3.5

Note: Each value is average of two tests for compressive strength and three tests for flexural strength.

[a]   Reference: 100% normal Portland cement. Control: 70% normal Portland cement plus 30% fly ash. Silica fume: 70% normal Portland cement plus 30% fly ash plus additions of condensed silica fume.
[b]   W/(C + FA) by weight.

charge through the concrete toward the reinforcing steel maintained at a positive potential. The plots of test results in Figure 12 show superior performance of condensed silica fume concrete in comparison with latex-modified or low-slump dense concrete.[7] These results suggest that the corrosion potential of silica fume concrete would be less than that of normal concrete. Factors that are responsible for this difference are lower permeability, limited mobility of chloride ions, and reduced oxygen diffusion in such concrete.

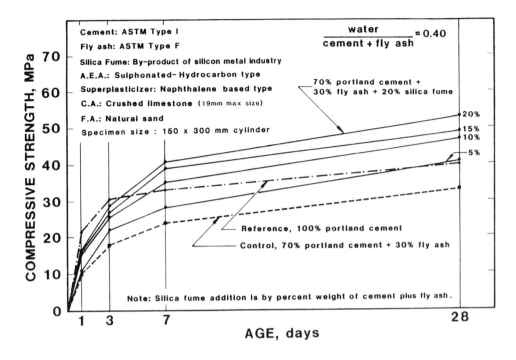

FIGURE 5.   Compressive strength vs. age for concrete with W/(C + F) of 0.40.[4]

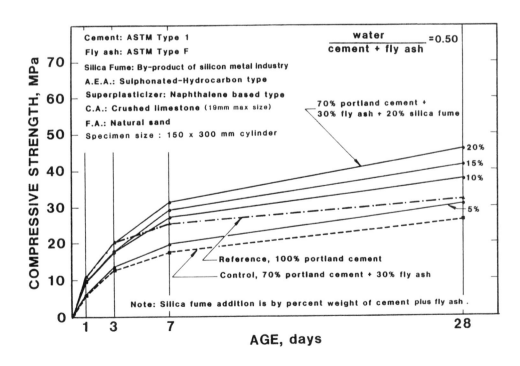

FIGURE 6.   Compressive strength vs. age for concrete with W/(C + F) of 0.50.[4]

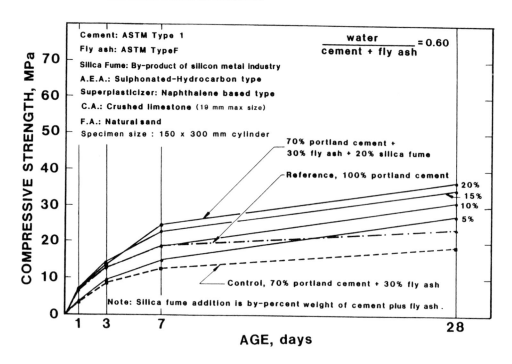

FIGURE 7.   Compressive strength vs. age for concrete with W/(C + F) of 0.60.[4]

Marusin[8] investigated the chloride ion penetration in conventional concrete and concrete containing condensed silica fume. The chloride ion penetration characteristics were studied on 100-mm concrete cubes, which were immersed in a 15% NaCl solution for 21 days. Following the 21-day soaking period and a subsequent 21-day air drying period, concrete samples were removed by drilling at depth intervals of 0 to 12, 12 to 25, 25 to 37, and 37 to 50 mm, and tested for acid-soluble chloride ion content using a potentiometric titration procedure.[8]

The properties of concrete samples investigated are shown in Table 10, and acid soluble chloride ion content is given in Tables 11 and 12. The highlights of Marusin's conclusions based upon these results are

1. Regardless of the percentage of condensed silica fume used, the top concrete layer (0 to 12 mm) of the samples tested contained similar amounts of chloride.
2. The best performance in terms of chloride ion content was shown by concrete incorporating 10% condensed silica fume. The chloride ion content at a depth of 12 to 25 mm attained the acid soluble corrosion threshold level of about 0.03% by weight of concrete, normally accepted for reinforced concrete, and was lower than this value at depths greater than 25 mm.
3. Of all the silica-fume concretes, the poorest performance in terms of chloride ion content was exhibited by the concrete with 5% condensed silica fume.

From these tests it appears that at silica fume contents beyond 10% the permeability for Cl⁻ ions increased. It is possible that as the percentage of silica fume is increased, increased amounts of silica fume remain uncombined and do not contribute to the lowering of permeability.

Table 7
MIXTURE PROPORTIONS OF CONCRETE CONTAINING BLAST-FURNACE SLAG, CONDENSED SILICA FUME, AND A SUPERPLASTICIZER[5]

Mix no.	Type of mixture[a]	W/(C + BFS)[b]	Relative proportions of cement and BFS (% by wt)		Condensed silica fume (% by wt of cement plus BFS)	Batch quantities (kg/m³)					Superplasticizer (l/m³ of concrete)
			Cement	BFS		Cement	BFS	Condensed silica fume	Fine agg.	Coarse agg.	
1	Reference		100	0	0	420	0	0	790	1055	—
2	Control		50	50	0	210	210	0	780	1060	—
3	5% silica fume	0.40	50	50	5	210	210	21	750	1055	0.8
4	10%		50	50	10	210	210	42	725	1055	2.1
5	15%		50	50	15	210	210	63	700	1055	2.6
6	20%		50	50	20	210	210	84	675	1050	3.6
7	Reference		100	0	0	330	0	0	870	1050	—
8	Control		50	50	0	165	165	0	865	1060	—
9	5% silica fume	0.50	50	50	5	165	165	17	845	1055	0.4
10	10%		50	50	10	165	165	33	825	1060	0.8
11	15%		50	50	15	165	165	50	805	1055	1.8
12	20%		50	50	20	165	165	66	785	1055	2.5
13	Reference		100	0	0	260	0	0	920	1050	—
14	Control		50	50	0	130	130	0	915	1060	—
15	5% silica fume	0.65	50	50	5	130	130	13	900	1055	0.2
16	10%		50	50	10	130	130	26	880	1055	0.5
17	15%		50	50	15	130	130	39	870	1055	1.2
18	20%		50	50	20	130	130	52	850	1055	1.6

[a] Reference mixture: 100% normal Portland cement, control mixture: 50% normal Portland cement plus 50% BFS, silica fume mixture: 50% normal Portland cement plus 50% BFS plus additions of condensed silica fume.

[b] W/(C + BFS) by weight.

Table 8

PROPERTIES OF FRESH CONCRETE CONTAINING BLAST-
FURNACE SLAG, CONDENSED SILICA FUME, AND A
SUPERPLASTICIZER[5]

Mix no.	Type of mixture[a]	W/(C + BFS)[b]	Properties of fresh concrete			
			Temp. (°C)	Slump (mm)	Unit wt (kg/m³)	Air content (%)
1	Reference		21	75	2435	1.2
2	Control		21	70	2430	1.1
3	5% silica fume	0.40	21	80	2415	1.3
4	10%		21	75	2410	1.3
5	15%		21	80	2410	1.3
6	20%		21	80	2400	1.2
7	Reference		21	75	2415	1.2
8	Control		21	70	2420	1.2
9	5% silica fume	0.50	21	75	2410	1.2
10	10%		21	70	2415	1.2
11	15%		21	70	2405	1.3
12	20%		21	80	2400	1.2
13	Reference		23	85	2400	1.3
14	Control		22	80	2405	1.1
15	5% silica fume	0.65	22	85	2400	1.2
16	10%		20	75	2390	1.1
17	15%		20	80	2395	1.3
18	20%		20	80	2385	1.3

[a] Reference mixture: 100% normal Portland cement, control mixture: 50% normal Portland cement plus 50% BFS, silica fume mixture: 50% normal Portland cement plus 50% BFS plus additions of silica fume.
[b] W/(C + BFS) by weight.

V. HIGH-STRENGTH NORMAL WEIGHT CONCRETE

In ordinary concrete the strength, primarily a function of the cementitious binder system, is usually low when compared with the strength of coarse aggregate. To obtain very high strengths, it is necessary to have a high-strength cement binder system together with relatively strong aggregates. This can be achieved by using a combination of condensed silica fume, Portland cement, and superplasticizers.[10-12] According to Bache,[10] the very high strength of the cement/silica fume/superplasticizer binder system is caused by dense packing. The principle of dense packing is that the ultrafine silica fume particles are densely packed in the spaces between the cement particles and normally distributed coarse and fine aggregates. Unlike the very fine cement particles which are angular in shape, the silica particles formed by condensation from gas phase are spherical, thus making the particles ideal for dense packing. The structures of the cement paste in fresh concrete, paste with a superplasticizer, and that with silica fume and a superplasticizer are shown in Figure 13. Compressive strengths of the order of 100 MPa and higher have been produced by judicious proportioning of normal Portland cement, condensed silica fume, and superplasticizers.[10] In these types of concretes, the use of high dosages of superplasticizers becomes mandatory in order to reduce water demand caused by the high percentages of silica fume. Strength data of silica-fume concrete are shown in Figure 14; concrete containing about 33% silica fume develops strength higher than 100 MPa within 1 month. Rupture surfaces in mortar

Table 9

COMPRESSIVE STRENGTH OF CONCRETE CONTAINING BLAST-FURNACE SLAG, CONDENSED SILICA FUME, AND A SUPERPLASTICIZER[5]

Mix no.	Type of mixture[b]	W/(C + BFS)[a]	Compressive strength of 100 × 200-mm cylinders (MPa)						
			1-day	3-day	7-day	28-day	56-day	91-day	180-day
1	Reference		20.6	32.7	38.3	47.1	52.3	53.9	59.3
2	Control		7.0	16.1	25.7	46.7	54.8	56.4	60.7
3	5% silica fume		7.0	17.6	30.3	57.1	62.4	64.0	66.9
4	10%	73.0	0.40	8.9	19.3	34.3	59.3	64.5	63.6
5	15%		9.1	21.9	34.5	60.7	65.2	65.1	69.4
6	20%		10.4	23.3	37.6	60.1	69.0	68.2	73.3
7	Reference		14.2	23.1	30.4	36.5	39.3	43.3	46.0
8	Control		5.0	12.7	18.8	33.0	37.2	42.4	47.9
9	5% silica fume	52.4	0.50	5.5	12.0	20.7	38.9	45.7	46.6
10	10%		5.3	14.7	25.1	45.5	51.8	52.6	54.2
11	15%		6.5	15.1	28.5	48.8	54.7	56.2	59.6
12	20%		6.8	15.3	29.3	47.5	54.8	56.6	61.9
13	Reference		8.7	16.0	21.0	28.3	30.5	32.5	33.7
14	Control		2.1	6.0	10.0	20.6	24.1	27.6	28.9
15	5% silica fume		2.5	6.7	11.9	26.2	31.3	35.2	34.8
16	10%	38.7	0.65	2.7	8.0	15.7	31.7	37.8	40.1
17	15%		3.4	8.3	18.5	35.3	41.3	45.2	45.1
18	20%		3.4	9.1	22.7	38.0	42.1	47.4	47.8

Note: Each values is average of three tests.

a W/(C + BFS) by weight.
b Reference mixture: 100% normal Portland cement, control mixture: 50% normal Portland cement plus 50% BFS, silica fume mixture: 50% normal Portland cement plus 50% BFS plus additions of silica fume.

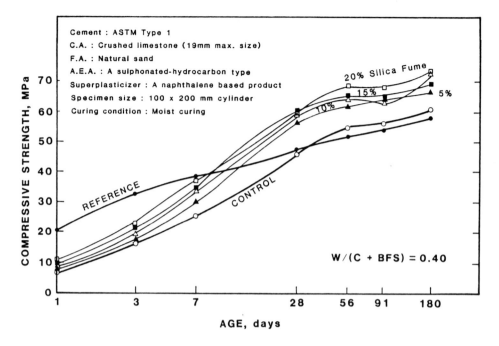

FIGURE 8.   Compressive strength vs. age for concrete with W/(C + BFS) of 0.40.[5]

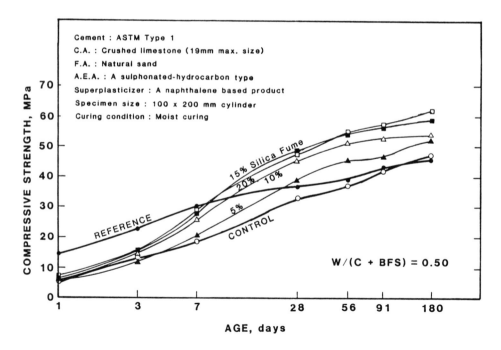

FIGURE 9.   Compressive strength vs. age for concrete with W/(C + BFS) of 0.50.[5]

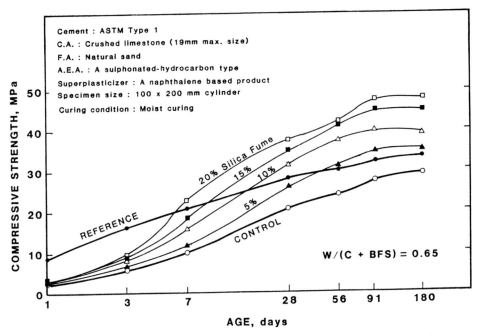

FIGURE 10. Compressive strength vs. age for concrete with W/(C + BFS) of 0.65.[5]

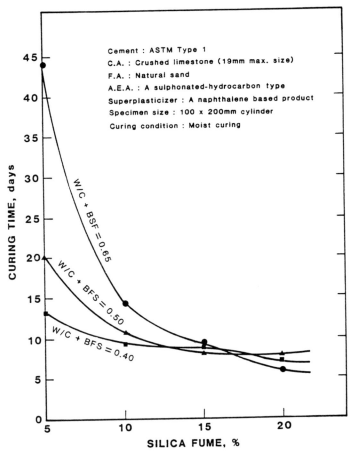

FIGURE 11. Time of curing required for silica fume concrete to reach compressive strength of reference concrete.[5]

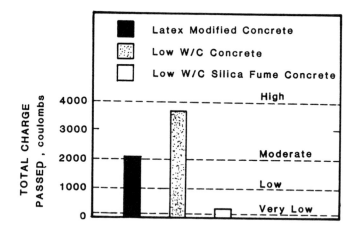

FIGURE 12.   Results of rapid chloride permeability test for various types of concrete.[7]

Table 10
MIX PROPORTIONS AND PROPERTIES OF FRESH AND HARDENED CONCRETE FOR CHLORIDE ION PENETRATION TEST[8]

Mix no.	Fine agg.	Coarse agg.	Silica fume	Portland cement	W/C (by wt)	Slump (mm)	Temp. (°C)	Air content (%)	Unit wt (kg/m³)	28-day compressive strength (MPa)
			Mix proportions (kg/m³)				**Properties of fresh concrete**			
CP1	645	1053	—	398	0.40	102	21	7.5	2256	35.3
CP2	660	1073	21	385	0.41	71	21	5.6	2307	46.5
CP3	638	1041	39	357	0.41	81	21	7.4	2236	41.7
CP4	643	1048	60	340	0.41	127	21	7.0	2256	48.8
CP5	643	1051	121	285	0.43	127	21	5.1	2275	54.1

Table 11
ACID-SOLUBLE CHLORIDE ION CONTENT[8]

Mix no.	Amt. of CSF (%)	0-12	12-25	25-37	37-50
		Chloride ion[a] (% by wt of concrete), for depth interval (mm)			
CP1	0	0.384	0.134	0.037	0.031
CP2	5	0.363	0.056	0.023	0.019
CP3	10	0.338	0.033	0.020	0.019
CP4	15	0.348	0.035	0.027	0.020
CP5	30	0.365	0.039	0.027	0.020

[a]   Average value of two samples.

191

Table 12
RELATIVE CHLORIDE ION
CONCENTRATIONS AT VARIOUS
DEPTHS[8]

Mix no.	Amt. of CSF (%)	Relative values (%)			
		0-12[a] mm	12-25 mm	25-37 mm	37-50 mm
CP1	0	100	34.9	9.6	8.1
CP2	5	100	15.4	6.3	5.2
CP3	10	100	9.8	5.9	5.6
CP4	15	100	10.1	7.8	5.7
CP5	30	100	10.7	7.4	5.5

[a] Chloride ion content within sample at depth interval from 0-12 mm is assumed to be 100%.

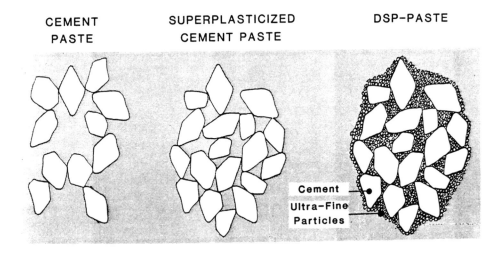

FIGURE 13. The structure of the paste in fresh concrete based on Portland cement, Portland cement + superplasticizer and Portland cement + silica fume + superplasticizer.[10]

and concrete made with ordinary cement paste and with silica fume are shown in Figure 15.

Wolsiefer[13] has reported data on the mechanical properties of concrete with a W/C (water to cementitious materials ratio) of 0.22. The mix proportioning and strength data are shown in Table 13. The cement factor used was 593 kg/m³ and 20% condensed silica fume was incorporated into the concrete. As can be seen from Table 13, the most significant compressive strength results were 65 MPa at 1 day and 124 MPa at 128 days. Wolsiefer has achieved 73 MPa at 1 day and 142 MPa at 28 days with the above type of concrete having lower W/C.

VI. ULTRA HIGH-STRENGTH CONCRETE

Ultra high-strength concrete may be classified as concrete having strength >150 MPa. Using principles of dense packing, Bache[10] has reported the development of the

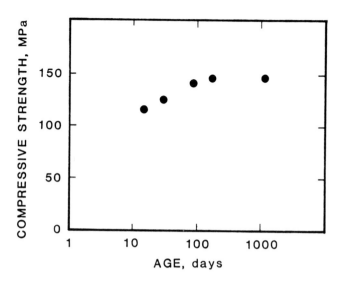

FIGURE 14.   Compressive strength of DSP-concrete cured in water at 20°C.[10] Silica fume, 133 kg/m³; Portland cement, 400 kg/m³; quartz sand $^1/_4$-1 mm, 141 kg/m³; quartz sand 1-4 mm, 566 kg/m³; crushed granite 8-16 mm, 1153 kg/m³; naphthalene-based superplasticizer (powder), 13.5 kg/m³; water, 100 kg/m³; consistency of fresh concrete: soft; and compaction: vibration for 10-20 sec at 50 Hz.[10]

above type of concrete. Compressive strengths ranging from 168.1 to 268.3 MPa for concretes and mortars, respectively, have been obtained using specialized small maximum size aggregates and high-curing temperatures (Table 14). This type of concrete is recommended for specialized applications and is not intended for routine concrete operations.

VII. HIGH-STRENGTH LIGHTWEIGHT CONCRETE

Conventional semi-lightweight concrete has been used to advantage in high-rise construction, especially for the structural floor systems. In this type of concrete, coarse aggregate consists of expanded shale or clay and fine aggregate is natural sand. The unit weight of this type of concrete is generally between 1840 to 1920 kg/m³ and the 28-day compressive strength is of the order of 35 to 40 MPa measured on 150 × 300-mm cylinders. In recent years the demand for high-strength lightweight concrete has been increasing because of the increasing heights of structures in the U.S., and thus the need to reduce weight. Also, the need for oil exploration and drilling platforms in the shallow waters of the Beaufort Sea in the Arctic region in Canada has led to the demand for high-strength semi-lightweight concrete. For example, in 1984, semi-lightweight high-strength concrete incorporating silica fume was used in the construction of a concrete island drilling system (CIDS).

Through judicious selection of mix proportions incorporating condensed silica fume, fly ash, and superplasticizers, it has become possible to produce air-entrained semi-lightweight concrete meeting the density requirements of 1500 to 2000 kg/m³ and reaching strengths of the order of 60 to 70 MPa at 28 to 90 days. Typical examples of mix proportions for the above type of concrete are given in Tables 15 to 17, together with the data for fresh and hardened concrete.[14] It is seen from the tables that both nonair- and air-entrained concretes reach values approaching 70 MPa. The lightweight aggregate used was expanded shale with moderate absorption characteristics. The high early-age strengths reported are probably the result of filler and pozzolanic action of condensed silica fume with fly ash contributing to later-age strength. It is believed that chemical/pozzolanic reaction at the interface of aggregate particles and cement/fly ash/condensed silica fume matrix also contribute to the stronger aggregate/matrix bond and strength. Notwithstanding this it is the crushing strength of the lightweight aggregate which will govern the required ultimate strength.

Data have also been reported from Switzerland on the high-strength lightweight concrete incorporating condensed silica fume and superplasticizers. According to Bürge,[15] by using expanded clay as a lightweight aggregate in combination with artificially introduced air voids in concrete, it was possible to obtain an air content of up to 40% of concrete volume and density/compressive strength values ranging from 1.1 t/m³/10 MPa to 1.8 t/m³/60 MPa. The strength data reported by Bürge were obtained on 120-mm cubes using DIN standards.

VIII. STEEL FIBER-REINFORCED CONCRETE

One major drawback of high-strength concrete with or without condensed silica fume is its lack of ductility. Such concretes, when subjected to sudden applied loads such as an earthquake or blast, will exhibit a sudden or catastrophic failure. It is well established that the incorporation of fibers in concrete increases its ductility, energy absorption, and ultimate strain capacity.[16] Carette et al.[14] have published limited data on semi-lightweight concrete incorporating steel fibers (Tables 15 to 17). Ramakrishnan and Srinivasan[17] have also reported data on normal weight concrete incorporating steel fibers which have deformed ends and are glued together in bundles with water-soluble adhesive (commercially known as Dramix fibers). The collating of fibers creates an artificial aspect ratio of approximately 30 when introduced to the concrete mix, and when the glue is dissolved by the mixing water, the fibers separate as individual fibers with an aspect ratio of 100.

The toughness index of fiber reinforced concrete is a measure of the amount of energy required to deflect a 102-mm prism by a given amount (1.9 mm) compared to the energy required to bring the prism to the point of first crack. It is calculated as the area under the load-deflection curve up to a deflection of 1.9 mm divided by the area under the load-deflection curve of the prism up to the first crack.[19] Toughness index values twice that of ordinary fiber reinforced concrete have been reported for some fiber reinforced silica fume concrete.[19] In general, the toughness index of fiber reinforced silica fume concrete is higher than the reference concrete without silica fume.[19]

IX. SLURRY-INFILTRATED FIBER CONCRETE (SIFCON)

According to Lankard,[20] slurry infiltrated fiber concrete can be rightly thought of as preplaced fiber concrete analogous to preplaced aggregate concrete. Fibers are placed in a form by hand or through the use of commercial fiber dispensing units. Light external vibration is applied during the fiber placement operation. The fiber loadings vary from 5 to 18% by volume, depending upon the fiber aspect ratio. Following this, cement slurry is poured onto the packed fiber bed. The infiltration of the slurry is assured by external vibration. After the slurry infiltration process is complete, the test specimens are cured in a normal manner.

Lankard's data show that SIFCON is a high-strength and highly ductile composite. SIFCON incorporating 5 to 18% of fibers by volume of concrete has compressive strengths of the order of 103 to 206 MPa and ultimate flexural strengths of the order of 27 to 69 MPa.

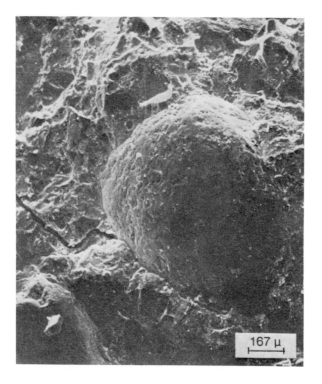

a

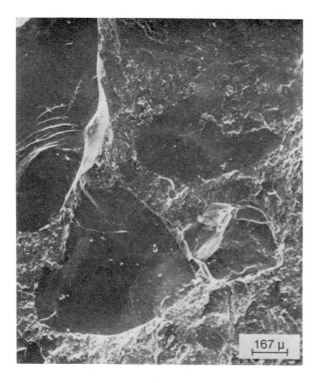

b

FIGURE 15.   Rupture surfaces in mortar and concrete made with ordinary cement paste (a and c) and with the new binder (b and d).

195

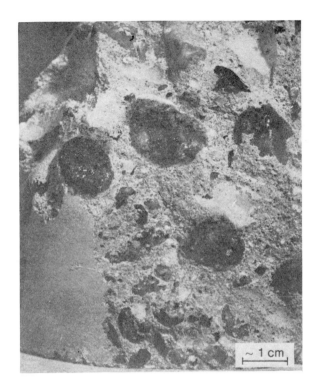

FIGURE 15c.

FIGURE 15d.

Table 13

MIX PROPORTIONS AND STRENGTH PROPERTIES OF
CONDENSED SILICA FUME ADMIXTURED CONCRETE[13]

Mix proportions & properties of fresh concrete	kg/m³	Compressive & flexural strengths at various time intervals		
		Age (days)	Comp. strength (MPa)	Flex strength (MPa)
		0.5	43.1	—
Type I cement	593	1	64.9	—
		3	77.9	11.1
Silica fume admixture	119	7	94.6	12.3
Elgin sand	537	14	102.8	—
		28	111.4	14.3
Thornton limestone	992	60	112.0	—
Water	158	90	115.1	—
		128	124.2	—
Av. slump	120 mm	365	126.5	—

Table 14

MECHANICAL PROPERTIES OF SOFT CAST DSP[a] MORTAR
AND CONCRETE[10]

Type of concrete or mortar (max. size aggregate)	Density (kg/m³)	Compressive strength at 28 days (MPa)	Sound velocity (m/sec)	Dynamic modulus of elasticity (GPa)
16 mm diabas	2666	168.1	4890	65
10 mm calcined bauxite	2878	217.5	6150	109
4 mm calcined bauxite	2857	268.3	6153	108

Note: Tests were done on 100 × 200 mm cylinders, cured in water at 60 to 80°C for 4 days.

[a]   Densified systems containing homogenously arranged, ultra-fine particles.

Balaguru and Kendzulak[21] have studied the mechanical properties of SIFCON concrete incorporating condensed silica fume. They have reported a significant increase in the flexural strength of SIFCON with silica fume as compared with that of the control mix.

Table 15

MIXTURE PROPORTIONS FOR HIGH-STRENGTH SEMI-LIGHTWEIGHT CONCRETE[14]

Mix no.	Batch quantities (kg/m^3)							Superplasticizer (% by wt of cementitious material)
	Cement	Fly ash	Silica fume	Fine aggregate	Coarse aggregate	Water	Fibers	
1	409	41	20	675	677	171	—	0.65
2	404	40	21	688	687	171	—	0.70
3	510	51	26	568	677	178	—	0.86
4	499	50	25	563	684	178	—	0.82
5	602	60	30	462	661	183	—	1.07
6	597	60	30	461	675	181	—	0.98
1	404	40	20	661	662	169	60	0.80
2	403	40	20	678	677	170	60	0.88
3	505	50	25	556	662	177	60	0.86
4	497	50	25	554	672	178	60	0.82
5	597	59	30	451	647	180	59	1.07
6	586	59	29	448	653	177	59	1.08

Note: Coarse aggregate: expanded shale with 19-mm maximum size, fine aggregate: natural sand, silica fume: more than 90% SiO_2 and surface area by BET = 20 m^2/g, superplasticizer: a naphthalene-based type, cement: ASTM Type 1, fly ash: low-calcium type, and steel fibers: Dramix ZP 30/0.5 with hooked ends.

Table 16

PROPERTIES OF FRESH CONCRETE FOR HIGH-STRENGTH SEMI-LIGHTWEIGHT CONCRETE[14]

	Mix no.	Initial slump (mm)	Slump after addition of superplasticizer (mm)	Inverted slump cone time (sec)	Density (kg/m^3)	Entrapped air (%)
With fibers	1	30	120	—	1992	2.1
	2	25	100	—	2011	2.2
	3	10	210	—	2011	1.6
	4	5	165	—	1998	1.8
	5	5	190	—	1998	1.4
	6	0	215	—	2004	2.4
Without fibers	1	5	85	19.0	2017	1.8
	2	5	70	22.0	2049	2.4
	3	0	90	12.0	2036	2.5
	4	0	85	18.0	2036	3.0
	5	0	150	11.5	2024	2.2
	6	0	165	14.0	2011	2.8

Table 17

PROPERTIES OF HARDENED CONCRETE FOR HIGH-STRENGTH SEMI-LIGHTWEIGHT CONCRETE[14]

Mix no.	Compressive strength (MPa)			Splitting tensile strength (MPa)		Flexural strength (MPa)			Toughness index			Impact strength[a] (no. of blows)	
	7 days	28 days	91 days	7 days	28 days	7 days	14 days	28 days	7 days	14 days	28 days	Sawn	Cast
1	41.8	52.0	—	2.8	4.4	5.3	6.6	6.6	—	—	—	—	—
2	40.0	55.2	61.5	3.7	4.7	6.2	6.3	6.9	—	—	—	90/98	85/97
3	49.8	60.1	—	3.9	5.3	6.0	7.4	7.8	—	—	—	—	—
4	51.7	61.8	64.5	4.2	5.4	6.4	7.8	7.8	—	—	—	145/150	80/85
5	55.2	66.6	—	5.4	5.4	7.6	8.0	8.4	—	—	—	—	—
6	56.0	64.9	68.7	4.7	5.6	7.2	6.9	9.0	—	—	—	25/51	58/84
1	43.4	56.6	—	6.1	7.4	5.9	6.3	7.1	22.8	21.8	21.8	—	—
2	41.3	55.5	61.3	6.1	7.3	6.4	6.5	6.6	21.8	19.9	21.2	363/519	300/900
3	49.5	61.4	—	6.8	7.6	6.9	8.0	7.6	19.9	17.1	18.8	—	—
4	52.3	62.6	65.2	7.0	7.7	7.1	7.4	7.6	15.7	20.1	18.9	325/550	350/425
5	55.2	65.1	—	7.4	8.5	6.9	8.0	7.9	18.6	15.3	21.9	155/333	350/575
6	56.2	61.7	66.1	7.5	7.6	8.0	7.7	7.8	18.1	15.0	15.3		

[a] The first number is the number of blows required to cause first crack, and the second number is the number of blows required for ultimate failure.

REFERENCES

1. Skrastins, J. I. and Zoldners, N. G., Ready-Mixed Concrete Incorporating Condensed Silica Fume, ACI Spec. Publ. Malhotra, V. M., Ed., American Concrete Institute, Detroit, 1983, 813.
2. Berry, E. E. and Malhotra, V. M., Fly Ash in Concrete, CANMET Spec. Publ. SP85-3, Energy, Mines and Resources Canada, Ottawa, 1985.
3. Mehta, P. K. and Gjorv, O. E., Properties of Portland cement concrete containing fly ash and condensed silica fume, *Cement Concrete Res.*, 12(5), 587, 1982.
4. Carette, G. G. and Malhotra, V. M., Early-Age Strength Development of Concrete Incorporating Fly Ash and Condensed Silica Fume, ACI Spec. Publ. SP 79, Vol. 2, Malhotra, V. M., Ed., American Concrete Institute, Detroit, 1983, 765.
5. Malhotra, V. M., Carette, G. G., and Aitcin, P. C., Mechanical properties of Portland cement concrete incorporating blast-furnace slag and condensed silica fume, in Proc. RILEM-ACI Symp. Technol. Concrete when Pozzolans, Slags and Chemical Admixtures are Used, Monterrey, Mexico, 1985, 395.
6. Whiting, D., *In-Situ* Measurements of the Permeability of Concrete to Chloride Ions, ACI Spec. Publ. SP 82, Malhotra, V. M., Ed., American Concrete Institute, Detroit, 1984, 501.
7. Christensen, D. W., Sorensen, E. V., and Radjy, F. F., Rockbond: a new microsilica concrete bridge deck overlay material, in Proc. Int. Bridge Conf., Pittsburgh, June 4-6, 1984 (copies of the paper available from Elkem Chemicals Inc., Parkwest Office Centre, Pittsburgh, Pa., 15275).
8. Marusin, S. L., Chloride ion penetration in conventional concrete and concrete containing condensed silica fume, Proc. 2nd Int. Conf. on the Use of Fly Ash, Silica Fume, Slag and Natural Pozzolans in Concrete, Malhotra, V. M., Ed., Madrid, April 1986.
9. Pfeifer, D. W. and Scali, M. J., Concrete Sealers for Protection of Bridge Structures, NCHRP Rep. No. 244, Transportation Research Board, Washington, D.C., December 1981.
10. Bache, H. H., Densified cement — ultra fine particle based materials, paper presented 2nd Int. Conf. on Superplasticizers in Concrete, Ottawa, Canada, 1981 (copies may be obtained from Aalborg Portland Cement, Aalborg, Denmark).
11. Malhotra, V. M., Ed., Superplasticizers in Concrete, ACI Spec. Publ. SP 62, American Concrete Institute, Detroit, 1978.
12. Malhotra, V. M., Ed., Development in Superplasticizers, ACI Spec. Publ. SP 68, American Concrete Institute, Detroit, 1981.
13. Wolsiefer, J., Ultra high strength field placeable concrete with silica fume admixture, *ACI Concrete Int. Design Constr.*, 6(4), 25, 1984.
14. Carette, G. G., Wilson, H. S., and Tan, C. W., Development of Superplasticized Lightweight High-Strength Concrete Incorporating Steel Fibers, Silica Fume and Fly Ash, CANMET Div. Report MRP/MSL 84-81, Energy, Mines and Resources Canada, Ottawa, 1984.
15. Bürge, T. A., High-Strength Lightweight Concrete With Condensed Silica Fume, ACI Spec. Publ., SP 79, Vol. 2, Malhotra, V. M., Ed., American Concrete Institute, Detroit, 1983, 731.
16. Johnston, C. D., Fiber-reinforced concrete, in *Progress in Concrete Technology*, Malhotra, V. M., Ed., CANMET, Energy, Mines and Resources Canada, Ottawa, 1980, 451.
17. Ramakrishnan, V. and Srinivasan, V., Performance Characteristics of Fiber Reinforced Condensed Silica Fume Concretes, ACI Spec. Publ. SP 79, Vol. 2, Malhotra, V. M., Ed., American Concrete Institute, Detroit, 1983, 797.
18. ACI Committee 544, Measurement of Properties of Fiber Reinforced Concrete, ACI 544.2R-78, American Concrete Institute, Detroit, 1978.
19. Ramakrishnan, V., Brandshaug, T., Coyle, W. W., and Schrader, E. K., A comparative evaluation of concrete reinforced with straight steel fibers and fibers with deformed ends glued together into bundles, *ACI J. Proc.*, 77(3), 135, 1980.
20. Lankard, D. R., Slurry-infiltrated fiber concrete, *ACI Concrete Int. Design Constr.*, 6(12), 44, 1984.
21. Balaguru, P. and Kendzulak, J., Use of Silica Fume in Slurry-Infiltrated Concrete, in Proc. 2nd Int. Conf. on the Use of Fly Ash, Silica Fume, Slag and Natural Pozzolans in Concrete, Malhotra, V. M., Ed., Madrid, April 1986.

Chapter 11

STANDARDS

I. INTRODUCTION

Generally there is a considerable time lag between the introduction of a new material in the marketplace and the issuance of a national or an international specification. Condensed silica fume is no exception. Even though the fume has been in use in concrete for the past 10 years in Norway, at present, there is no comprehensive Norwegian specification, though a considerable body of experience has been developed in that country and an upper limit of 8% has been introduced for the use of silica fume in concrete. In North America, though the silica fume was introduced only 5 years ago, there has been considerable progress in the development of specifications. The probable reason for this rapid progress is that both Canada and the U.S. already had specifications for pozzolanic materials such as fly ash, and it was thus relatively easy to accomodate a new highly pozzolanic material by minor but significant additions to the existing specifications. In Canada, the clauses dealing with physical and chemical requirements and guidelines for the use of silica fume in concrete were approved in 1985 by the Canadian Standards Association Committee on Concrete, and are being processed for publication in 1987. In the U.S., a draft standard has been developed by the ASTM, and in all likelihood it will be issued in early 1988. South Africa is another country where some progress is being made toward the development of a specification for the use of condensed silica fume in concrete.

II. NORWEGIAN LIMITS ON THE USE OF SILICA FUME IN CONCRETE

As already mentioned, there is no comprehensive Norwegian specification on condensed silica fume as a material. However, the following limits have been introduced in existing standards to cover its use in concrete:[1]

1. Silica fume with at least 85% SiO_2 can be added to concrete under certain conditions and the amount of fume cannot exceed 8% by weight of cement.
2. When using silica fume, the amount of cement must be at least 240 kg/m³ and the W/(C + SF) must not exceed 0.70.
3. The errors in batching silica fume must be less than 5%.

III. CANADIAN SPECIFICATIONS FOR SILICA FUME

The Canadian Standards Association is the leading specification writing organization in Canada, and the Standard A23.5 "Supplementary Cementing Materials and Guidelines for their Use in Concrete" covers the requirements for the use of fly ash, slag, and natural pozzolans in concrete.[2] The new edition of the above standard to be issued in 1987 will have clauses dealing with condensed silica fume. In this standard the condensed silica fume is defined as a finely divided residue resulting from the production of silicon or silicon-containing alloys and which is carried from the burning area of a furnace by exhaust gases. Only the fume resulting from the production of silicon or ferrosilicon alloys containing at least 75% silicon[3] are satisfactory for use in concrete.

Table 1
CHEMICAL REQUIREMENTS[3]

Property	Material type			
	N	F	C	U (Silica fume)
SiO_2, min (%)	—	—	—	85
SO_3, max (%)[a]	3.0	5.0	5.0	1.0
Loss on ignition max (%)	10.0	12.0	6.0	6.0

Note: N refers to natural pozzolans, F, low-calcium fly ashes, and C, high-calcium fly ashes.

[a] This limit may be exceeded, provided that the supplementary cementing material, when tested in combination with the particular Portland cement with which it is to be used, exhibits expansion not in excess of 0.020% at 14 days when tested in accordance with Clause 7.5.5 of CAN3-A5. In the text mix, replace a mass of Portland cement by an equal mass of supplementary cementing material in the amount of 20% generally or for Type U in the amount of 10% or the anticipated maximum field replacement percentage, whichever is greater.

A. Chemical Requirements

The chemical requirements covering SiO_2, SO_3, and loss on ignition (LOI) as approved by the Canadian Standard[2] are shown in Table 1 together with the limits for the natural pozzolans and fly ashes. According to Isabelle,[3] the rationale for the selection of limits shown in Table 1 is as follows.

The most important of these requirements is the minimum SiO_2 content which has been placed at 85% because it has been found that silica fume having at least 85% SiO_2 generally performs well whereas that having a lesser amount of SiO_2 does not perform as well. Furthermore, the silica fumes from Canadian sources have SiO_2 content well in excess of the specified minimum.

Silica fume contains very little SO_3, principally because of the low sulfur carbon used in the manufacturing process. This is done on purpose in order to obtain silicon and ferro-silicon alloys of the highest quality. In view of the above, the Subcommittee debated the need to put any limit at all on SO_3 but finally allowed for a maximum of 1%.

Loss on ignition has been set at 6% maximum although in practice it is generally much lower. However, when wood chips are used in the manufacture of silicon or ferro-silicon alloys, some of the fine wood chips may find their way into the fume causing an increase in the loss on ignition; they do not adversely affect its intrinsic quality.

B. Physical Requirements

As in the existing specifications for fly ash, the physical requirements have been divided into two separate parts, one covering the mandatory and the other covering the optional requirements.

The mandatory requirements covering accelerated pozzolanic activity index, fineness, soundness, and uniformity are shown in Table 2. The nonmandatory or optional requirements governing drying shrinkage, air entraining admixture dosage, and reactivity with cement alkalis are given in Table 3.

The inclusion of the accelerated pozzolanic activity index requirement is premature because of the lack of sufficient published data on the test method. The fineness measurement by determining the percentage retained on a 45-μm sieve is also arbitrary. The

Table 2
MANDATORY PHYSICAL REQUIREMENTS[3]

Property	Material type			
	N	F	C	U (Silica fume)
Accelerated pozzolanic activity index, min, % of control	68	68	68	85
Fineness, max, % retained on 45-μm sieve (wet sieving)	34	34	34	10
Uniformity requirements:				
soundness — autoclave expansion or contraction (%)	0.8	0.8	0.8	0.2
Relative density, max variation from av. (%)	5	5	5	5
Fineness, max variation from average, % points	5	5	5	5

Note: The accelerated pozzolanic activity index determination is performed by carrying out the test at 65°C instead of 38°C as for the normal pozzolanic index determination.

Table 3
OPTIONAL PHYSICAL REQUIREMENTS FOR POZZOLANS[3]

Property	Material type			
	N	F	C	U (Silica fume)
Increase of drying shrinkage, max % points of control	0.03	0.03	0.03	0.03
Uniformity of addition rate of an air entraining agent: max variation from average (%)	20	20	20	20
Reactivity with cement alkalis: min reduction (%)	75	60	60	80

correct way to determine fineness is by the BET nitrogen adsorption technique;[4] the standard should have included this technique. The uniformity and density requirements are the same as those for natural pozzolans and fly ashes in the existing standard.

The maximum allowable autoclave expansions have been set at 0.2%. The rationale for the selection of a value of 0.2% is based on the premise that the limited amount of data available to the Committee indicated values much lower than the one being adopted.

As for the nonmandatory physical requirements, the drying shrinkage and uniformity of the addition rate of an air-entraining admixture are the same as those for natural pozzolans. The requirement that condensed silica fume must reduce mortar expansion due to the reaction between alkalis in cement and reactive silica in aggregates at 14 days by at least 80% as compared with control is somewhat arbitrary because no detailed published data are available. The probable rationale is that because silica fume contains a high percentage of amorphous silica, it should be more effective in controlling

Table 4
METHODS OF TEST FOR POZZOLANS[3]

Reference standards

Chemical tests	
Sulfur trioxide (SO₃)	CAN3-A5
Loss on ignition (LOI)	ASTM C311
Silica (SiO₂)	ASTM C114
Physical tests	
Pozzolanic activity index, 7 days	ASTM C311 (Sect. 29 to 32) except store both control and test cubes 6 days at 65°C and test at 7 days
Drying shrinkage	ASTM C311 (Sect. 22 to 32) except that for testing silica fume, use 500 g of Portland cement, 50 g of silica fume, and 1325 g of sand in the test mix
Relative density	ASTM C311 (Sect. 20)
Soundness	ASTM C311 (Sect. 25)
Reactivity with cement alkalis	ASTM C441
Uniformity of air content	ASTM C311 (Sect. 27 & 28)
Fineness: wet sieved on 45-μm sieve	CAN3-A5

the alkali-silica reaction in concrete. Some evidence suggests that cement containing silica fume adsorbs more alkali and alkaline earth cations than the control cement containing no silica fume.

C. Methods of Test

The testing methods for ensuring that condensed silica fume meets the specification requirements are given in Table 4. These are primarily ASTM test methods,[4] except for the determination of SO₃ and the percentage retained on a 45-μm sieve for which reference is made to test methods given in Canadian Standard CAN3-A5.[5]

IV. U.S. SPECIFICATIONS

In the U.S., the ASTM is the leading specification writing organization, and the ASTM Designation C618 covers the specification for fly ash and raw or calcined natural pozzolan for use as a mineral admixture in Portland cement concrete.[4] The Subcommittee responsible for this specification is drafting a revised C618 which will broaden the scope of the above specification to allow the use of condensed silica fume. This specification is likely to follow very closely that of Canada and the clauses dealing with condensed silica fume will parallel, with minor exceptions, those which have been adopted in Canada. No attempt is being made to introduce the accelerated pozzolanic activity test and in all likelihood the air permeability methods will be recommended for surface area determination. This method is not totally satisfactory for a very fine material like condensed silica fume. It is expected that ASTM will issue the revised C618 sometimes in 1988.

V. CANADIAN STANDARDS FOR THE USE OF SILICA FUME IN CONCRETE

At present, concrete standards in Canada do not allow the use of silica fume in concrete. However, this situation may change with the issuance of a materials standard covering silica fume. The consensus in the Committee, which is drafting the relevant clauses for incorporation in the standard dealing with concrete, is that an upper limit of 10% should be placed for the use of silica fume in concrete. The rationale for limiting the use to 10% is the excessive drying shrinkage and high water demand of concrete if this percentage is exceeded. There is also some concern that higher dosages may have an adverse effect upon the frost resistance of concrete.

This Committee is also recommending that the water-cementing materials ratio of silica fume concrete may be increased by 0.01 for each percentage of silica fume added up to a maximum of 0.05. This was in recognition of the greater water demand of silica fume compared with other pozzolans. The Committee is also of the opinion that the increase in the water-cementing materials ratio is not detrimental to the durability of concrete.

VI. BLENDED PORTLAND/SILICA FUME CEMENT

A number of countries are producing blended cements incorporating silica fume; two are Iceland and Canada.[3] In Canada the CSA Standard CAN3-A362-M83, "Blended Hydraulic Cements" was amended in 1985 to broaden its scope to cover condensed silica fume.[6] It allows the use of up to 15% condensed silica fume in the blended product and is designated as Portland pozzolan cement type 10PM. This type of cement, being marketed in Canada in a very limited quantity, contains 6 to 8% condensed silica fume.

REFERENCES

1. Norwegian National Standards NS 3420 and NS 3474 (with amendments in 78-06-01), 1978.
2. CSA Standard A23.5, Supplementary Cementing Materials and Guidelines for Their Use in Concrete, Canadian Standards Association, Rexdale, Ontario (a revised edition including the clauses covering condensed silica fume was issued in 1986).
3. Isabelle, H. L., Development of a Canadian Specification For Silica Fume, Proc. 2nd Int. Conf. on the Use of Fly Ash, Silica Fume, Slag and Natural Pozzolans in Concrete, Malhotra, V. M., Ed., 1986, 1577.
4. Annual Book of ASTM Standards, Vol. 04.02, American Society for Testing and Materials, Philadelphia, 1983.
5. CSA Standard CAN3-A5-M83, Portland Cements, Canadian Standards Association, Rexdale, Ontario, 1983.
6. CSA Standard CAN3-A362-M83, Blended Hydraulic Cements, Canadian Standards Association, Rexdale, Ontario, 1983 (amended 1985).

Chapter 12

CONDENSED SILICA FUME IN CEMENT: BIOLOGICAL CONSIDERATIONS

I. INTRODUCTION

There is a large body of data relating to the biological effects of inhaled particulates. The well-known occupational hazards resulting from exposures to some crystalline forms of silica (fibrosis) and to silicates (fibrosis and cancerogenesis) refer mainly to occupationally related lung pathologies.

The mechanisms of action of respirable particulates are not fully understood. However, two sets of parameters appear to stand out from the experimental data: physical parameters, such as size and geometry, and chemical parameters such as surface charge and surface chemistry. It is now generally recognized that both sets of parameters contribute to the total biological reactivity potential of a given particulate.

II. PHYSICAL PARAMETERS

It is obvious that particulates which may impact on pulmonary functions and integrity must be airborne. Furthermore, in order to reach deep into the pulmonary lobes, where the terminal bronchioles end anatomically in microscopic dead ends (alveoli) in which gas exchanges take place, and where specialized cells (macrophages) perform their scavenging function (phagocytosis), the airborne particles must be "respirable". Therefore, large airborne particles which are respirable are normally filtered by physical barriers (nasal hairs), and/or are physically trapped by mucous and entrained up the respiratory tree by ciliary movements.

Smaller airborne particles which are respirable may escape those physical barriers and eventually reach deep into the respiratory tree. At one end of the size spectrum, the particles may be small enough to be engulfed by macrophages (phagocytosis). At the other end of the spectrum, some respirable particles may be too large to undergo complete phagocytosis by a single macrophage. These larger particles are attacked by many macrophages. As a result, the cells may be damaged in the process due to incomplete phagocytosis, and may release fibrogenetic factors.

The situation is further complicated by the fact that some particulates may be quite long (up to 30 μm), but at the same time thin enough (diameter <2 μm) to allow them to reach deep into the alveoli. This is the special case of fibrous particles. Stanton and Layard[2] have studied the correlation between fiber size and biological reactivity (tumorigenicity), using materials of different chemical compositions. The strongest correlation found was with fibers 0.25 μm in diameter and 8.0 μm in length. Fibers studied in adjacent size ranges also showed correlation, although to a lesser degree.

Some recent data on the biological effects (genotoxicity) of fibrous erionite, whose size characteristics are much smaller than the critical range proposed by Stanton, show that this material is possibly the most powerfully tumorigenic particulate ever tested.[2,3] Thus, it appears that while important, the dimensions and geometry of particles are not the only parameters involved. Indeed, much recent evidence points to chemical characteristics of respirable particles as parameters of paramount importance in determining their biological reactivity.[4]

III. CHEMICAL PARAMETERS

Once inhaled, small respirable particles may be rapidly engulfed by scavenger cells, and disposed of by normal physiological pathways, without undue burden to normal defense mechanisms. For instance, there is growing evidence that exposure to short (L <2 μm) asbestos fibers will not lead to the observable effects seen with longer fibers.[5-8]

However, in some cases, even "short" fibers may present surface chemical characteristics which are potent enough to lead to pathological damage, as demonstrated by the powerful tumorigenic potential of fibrous erionite.

Nolan et al.[9] have shown that all silica particles are not equally biologically active, and that their membranolytic potential is related to "surface functionalities". These surface functionalities, which have been related to the silanol groups formed when the mineral is hydrated, can be modified by a variety of chemical treatments which will affect the biological reactivity potential of the particles without changing their size distribution.[10]

There are other indications that chemical surface characteristics also contribute to the overall pathogenic potential of respirable particles. For instance, some materials display very strong chemical affinity toward well-known organic carcinogens, such as some polyaromatic hydrocarbons (i.e., benzo-a-pyrene). Such materials could also act as carriers for carcinogenic agents, even if in and by themselves they show low cytotoxic activity. This parameter is therefore of great importance when dealing with respirable particles in workers who smoke, in that it may determine the potential degree of risk for lung carcinoma.

IV. CONDENSED SILICA FUME

The use of condensed silica fume (CSF) in concrete is rather recent, the first reported experiment dating back to the early 1950s.[11] Because of the widely recognized silica-related lung disease (silicosis), the question whether there is a risk resulting from exposure to CSF is justified.

In spite of differences in production processes, it seems that different conditions will all lead to the production of spherical particles, whose mean diameter is between 0.1 to 0.25 μm. For this reason, all varieties of CSF should be considered "respirable", i.e., able to reach down into the lung airways. Also for the same reason, all are likely to be completely engulfed by macrophages. This does not necessarily indicate that these small particles should be considered innocuous, as other equally small particles have been shown to be highly tumorigenic.

Because of the relatively recent use of CSF, information concerning possible biological effects in man is very scanty. These have been reviewed recently by Jahr.[12] While available reports indicate that CSF may not be devoid of biological activity, the responses observed are certainly not of the same degree of severity and permanence compared to free silica.

There are significant differences in the fibrogenic potential of different types of free silica, the severity and speed of fibrosis being greatest due to tridymite, then in descending order, to cristobalite, quartz, and least of all to vitreous silica. It is likely that what adverse health effects that may have been observed following exposure to CFS must be ascribed to small contamination of the amorphous silica with cristobalite and/ or tridymite, for instance in some industrial processes where the material is used in mixtures submitted to high temperature, such as in the production of refractory material. This phenomenon has been found to occur when relatively innocuous natural diatomite is calcined. When used as an agent in the production of high density cement, uncalcined CFS is unlikely to be contaminated by the crystalline fibrogenic forms of silica. At any rate, Jahr[12] has warned users of amorphous silica containing particles <5 μm that the material may be transformed to cristobalite upon heating. Thus, the need to monitor for the presence of these crystalline forms in presumed "amorphous" CSF. An additional reason for this caution is the report[13] of experiments showing that tumors can be induced in animals by intrapleural injections of tridymite, cristobalite, and crystalline silica.

Finally, one of the obvious questions which remains to be answered is the potential for CSF to adsorb and carry polyaromatic hydrocarbons. This should be verified in order to determine any potential for synergism of action with known environmental carcinogens which may be present in the workplace, and especially in exposed workers who smoke.

REFERENCES

1. Stanton, M. F. and Layard, M., Carcinogenicity of natural and man-made fibers, *Adv. Chem. Oncol.*, 1, 181, 1978.
2. Poole, A., Brown, R. C., Turner, C. J., Skidmore, J. W., and Griffith, D. M., In vitro genotoxic activities of fibrous erionite, *Br. J. Cancer*, 47, 697, 1983.
3. Wagner, J. C., Health hazards of substitutes, in World Symposium on Asbestos, Montreal, 1982, 244.
4. Dunnigan, J., Biological effects of fibers: Stanton's hypothesis revisited, *Environ. Health Perspect.*, 57, 333, 1984.
5. Davis, J. M. G., The fibrogenic effects of mineral dusts injected into the pleural cavity of mice, *Br. J. Exp. Pathol.*, 53, 190, 1972.
6. Lemaire, I., Characterization of bronchoalveolar cellular response in experimental asbestosis, *Am. Rev. Resp. Dis.*, 131, 144, 1985.
7. Platek, S. F., Groth, D. H., Ulrich, C. E., Stettler, L. E., Finnell, M. S., and Stoll, M., Chronic inhalation of short asbestos fibers, *Fundam. Appl. Toxicol.*, 5, 327, 1985.
8. Davis, J. M. G., Addison, J., Bolton, R. E., Donaldson, K., Jones, A. D., and Smith, T., The pathogenicity of long versus short fiber samples of amosite asbestos administered to rats by inhalation and intraperitoneal injection, *Br. J. Exp. Pathol.*, 67, 415, 1986.
9. Nolan, R. P., Langer, A. M., Harington, J. S., Oster, G., and Selikoff, I. J., Quartz hemolysis as related to its surface functionalities, *Environ. Res.*, 26, 503, 1981.
10. Nolan, R. P., Langer, A. M., and Foster, K. W., Particle size and chemically-induced variability in the membralytic activity of quartz: preliminary observations, in *3rd Int. Workshop on the In Vitro Effects of Mineral Dusts*, NATO ASI Ser. G, Vol. 3, Beck, E. and Bignon, J., Eds., Schluchsee/Black Forest, West Germany, 1984, 39.
11. Bernhardt, C. J., SiO$_2$ Dust as Cement Additive, Betogen Idag, 17, ARG No. 2, April 1952.
12. Jahr, J., Possible health hazards from different types of amorphous silicas/suggested threshold limit values, in *Condensed Silica Fume,* Aitcin, P. C., Ed., University of Sherbrooke, Quebec, 1983, 28.
13. Wagner, M. M. F., Wagner, J. C., Davies, R., and Griffith, D. M., Silica-induced malignant histocytic lymphoma: incidence linked with strain of rat and type of silica, *Br. J. Cancer,* 41, 908, 1980.

ADDITIONAL REFERENCES

1. Sharp, J. W., Silica Modified Cement, U.S. Patent 2,410,954, 1946.
2. Newell, W. J. and Madden, J. E., Cement Plaster, U.S. Patent 3,135,617, 1964.
3. Schulze, H. C., Method for Extrusion, U.S. Patent 3,880,664, 1975.
4. Occleshaw, J. E. and Smith, T. E., Improvements In and Relating to Board Products, U.S. Patent 3,969,567, 1976.
5. Kjohl, O. and Olstad, P. H., Process for Manufacturing Concrete of High Corrosion Resistance, U.S. Patent 4,118,242, 1978.
6. Asgeirsson, H. and Gudmundsson, G., Pozzolanic activity in silica dust, *Cement Concrete Res.*, 9(2), 249, 1979.
7. Traetteberg, A., Silica Fume in Ready-Mixed Concrete — Evaluation of the Hydration Course, Paper No. STF65 F79014, Norwegian Cement and Concrete Research Institute, Trondheim, 1979.
8. Bache, H. H., *Cement Bound Materials with Extremely High Strength and Durability,* Aalborg Portland, Aalborg, Denmark, 1980.
9. Okkenhaug, K., Pozzolanas in concrete, *Nordisk Beton* 24(3), 27, 1980.
10. Dunnom, D. D. and Wagner, M. P., The classification of silicon dioxide powders, *ASTM Standard. News*, 10, 1981.
11. Gjorv, O. E. and Loland, K. E., Eds., Condensed silica fume in concrete, in Proc. Nordic Research Seminar on Condensed Silica Fume in Concrete, Trondheim, Norway, December 1981.
12. Loland, K. E., Silica Fume in Concrete, Paper No. STF65 F81011, Norwegian Cement and Concrete Research Institute, Trondheim, 1981.
13. North, J. W., Process of Using and Products from Amorphous Silica Particulates, U.S. Patent 4,297,309, 1981.
14. Simmons, D. D., Pasko, T. J., and Jones, W. R., Properties of Portland Cement Concretes Containing Pozzolanic Admixtures, Rep. No. FHWA/RD-80/184, U.S. Federal Highway Administration, Washington, D.C., 1981.
15. Aarsleff, L. and Molbak, K., Silica-concrete for structural application — some case studies, in *Condensed Silica Fume in Concrete,* Norwegian Institute of Technology, Trondheim, 1982, 265.
16. Cornwall, C. E., Compositions of Cementitious Mortar, Grout and Concrete, U.S. Patent 4,310,486, 1982.
17. Dahl, P. A., Plastic shrinkage of silica concrete — consequences and preventive measures, in *Condensed Silica Fume in Concrete,* Norwegian Institute of Technology, Trondheim, 1982, 245.
18. Gudmundsson, G., Production of blended cement based on silica fume — practical experiences, in *Condensed Silica Fume in Concrete,* Norwegian Institute of Technology, Trondheim, 1982, 135.
19. Hjorth, L., Microsilica in Concrete, Nordic Concrete Res. Publ. No. 1, Nordic Concrete Federation, Oslo, 1982, 9.1.
20. Hogstad, O., Experiences from ready-mixed production of silica concrete, in *Condensed Silica Fume in Concrete,* Norwegian Institute of Technology, Trondheim, 1982, 287.
21. Kompen, R., Practical experiences with silica concrete in the field, in *Condensed Silica Fume in Concrete,* Norwegian Institute of Technology, Trondheim, 1982, 253.
22. Lehtonen, V., Silica fume in concrete — a project description, in *Condensed Silica Fume in Concrete,* Norwegian Institute of Technology, Trondheim, 1982. 163.
23. Loland, K. E. and Gjorv, O. E., Condensed silica fume in concrete, in *Condensed Silica Fume in Concrete,* Norwegian Institute of Technology, Trondheim, 1982, 165.
24. Malhotra, V. M. and Carette, G. G., Silica fume — a pozzolan of new interest for use in some concretes, *Concrete Constr.*, 27(5), 443, 1982.
25. Meland, I., Heat of hydration for cement paste with silica fume, in *Condensed Silica Fume in Concrete,* Norwegian Institute of Technology, Trondheim, 1982, 105.
26. Modeer, M., Silica fume as replacement for some of the cement in ordinary concrete with standard Portland cement; compressive cube strength tests, in *Condensed Silica Fume in Concrete,* Norwegian Institute of Technology, Trondheim, 1982, 91.
27. Okkenhaug, K., The effect of W/C ratio on the evaluation of the pozzolanic activity of silica fume, in *Condensed Silica Fume in Concrete,* Norwegian Institute of Technology, Trondheim, 1982, 75.
28. Olafsson, H., Effect of silica fume on the alkali — silica reactivity of cement, in *Condensed Silica Fume in Concrete,* Norwegian Institute of Technology, Trondheim, 1982, 141.
29. Olstad, P. H., Combination of silica fume and different types of cement, in *Condensed Silica Fume in Concrete,* Norwegian Institute of Technology, Trondheim, 1982, 121.
30. Opsahl, O. A., Freeze-thaw resistance of silica concrete, in *Condensed Silica Fume in Concrete,* Norwegian Institute of Technology, Trondheim, 1982, 203.
31. Opsahl, O. A., Steel Fiber Reinforced Shotcrete for Rock Support, Paper No. NTNF 1053.09511, Royal Norwegian Council for Scientific and Industrial Research (NTNF), Oslo, 1982.

32. Palm, C., Silica fume in concrete with low water-cement ratios, in *Condensed Silica Fume in Concrete,* Norwegian Institute of Technology, Trondheim, 1982, 151.

33. Ramakrishnan, V. and Srinivasan, V., Silica fume in fiber reinforced concrete, *Indian Concrete J.,* 56(12), 326, 1982.

34. Rickne, S. and Svensson, C., The new Tjorn bridge, *Nordisk Betong,* 26(2—4), 213, 1982.

35. Samuelsson, P., The influence of silica fume on the risk of efflorescense on concrete surfaces, in *Condensed Silica Fume in Concrete,* Norwegian Institute of Technology, Trondheim, 1982, 235.

36. Sorensen, E. V., Concrete with condensed silica fume. A preliminary study of strength and permeability, in *Condensed Silica Fume in Concrete,* Norwegian Institute of Technology, Trondheim, 1982, 189.

37. Vennesland, O., Corrosion of steel embedded in silica concrete, in *Condensed Silica Fume in Concrete,* Norwegian Institute of Technology, Trondheim, 1982, 59.

38. Virtanen, J., Silica fume in concrete, in *Condensed Silica Fume in Concrete,* Norwegian Institute of Technology, Trondheim, 1982, 155.

39. Aitcin, P. C., Annotated bibliography of the English language literature on condensed silica fume, in *Condensed Silica Fume,* Aitcin, P. C., Ed., University of Sherbrooke, Sherbrooke, Canada, 1983, 34.

40. Aitcin, P. C., Ed., *Condensed Silica Fume,* University of Sherbrooke, Sherbrooke, Canada, 1983.

41. Aitcin, P. C., Historical data on the use of condensed silica fume in concrete, in *Condensed Silica Fume,* Aitcin, P. C., Ed., University of Sherbrooke, Sherbrooke, Canada, 1983.

42. Aitcin, P. C., Influence of condensed silica fume on the properties of fresh and hardened concrete, in *Condensed Silica Fume,* Aitcin, P. C., Ed., University of Sherbrooke, Sherbrooke, Canada, 1983, 25.

43. Aitcin, P. C., Physico-chemical characteristics of condensed silica fume, in *Condensed Silica Fume,* Aitcin, P. C., Ed., University of Sherbrooke, Sherbrooke, Canada, 1983, 16.

44. Aitcin, P. C., Terminology, in *Condensed Silica Fume,* Aitcin, P. C., Ed., University of Sherbrooke, Sherbrooke, Canada, 1983, 6.

45. Burge, T. A., 14,000 psi in 24 hours, *Concrete Int. Design Constr.,* 5(9), 36, 1983.

46. Chatterji, S., Collepardi, M., and Moriconi, G., Pozzolanic property of natural and synthetic pozzolans: a comparative study, ACI Spec. Publ. SP79, Vol. 1, Malhotra, V. M., Ed., American Concrete Institute, Detroit, 1983, 221.

47. Fischer, K. P., Bryhn, O., and Aagaard, P., Corrosion of Steel in Concrete: Some Fundamental Aspects of Concrete with Added Silica, Rep. No. 51304-06, Norwegian Geotechnical Institute, Oslo, 1983.

48. Gautefall, O. and Vennesland, O., Effect of Cracks on the Corrosion of Embedded Steel in Silica Concrete Compared to Ordinary Concrete, Nordic Concrete Res. Publ. No. 2, Nordic Concrete Federation, Oslo, 1983, 17.

49. Gram, H. E., *Durability of Natural Fibers in Concrete,* Swedish Cement and Concrete Research Institute, Stockholm, 1983.

50. Hansen, T. C. and Narud, H., Recycled concrete and silica fume make calcium silicate bricks, *Cement Concrete Res.,* 13(5), 626, 1983.

51. Hansson, E. S., Soepler, B., and Reichert, G., Manufacture of Modified Portland Cements With Fly Ash and Blast Furnace Slag in the Norwegian Cement Industry, Zement-Kalk-Gips (Translation ZKG No. 12/82), 1983, 28.

52. Holland, T. C., Abrasion-Erosion Evaluation of Concrete Mixtures for Stilling Basin Repairs, Kinzua Dam, Pennsylvania, WES Misc. Paper SL-83-16, U.S. Army Corps of Engineers Waterways Experiment Station, Vicksburg, Miss., 1983.

53. Holland, T. C., Abrasion-Erosion Update, *Concrete Struct. Repair Rehab.,* 83(1), 1, 1983.

54. Kohno, K., Horii, K., and Fukushima, H., Use of fly ash, blast-furnace slag and condensed silica fume for concrete block stripped immediately after molding, ACI Spec. Publ. SP79, Vol. 2, Malhotra, V. M., Ed., American Concrete Institute, Detroit, 1983, 1165.

55. Lessard, S., Aitcin, P. C., and Regourd, M., Development of a low heat of hydration blended cement, ACI Spec. Publ. SP79, Vol. 2, Malhotra, V. M., Ed., American Concrete Institute, Detroit, 1983, 747.

56. Malhotra, V. M., Condensed silica fume, an introduction, in *Condensed Silica Fume,* Aitcin, P. C., Ed., University of Sherbrooke, Sherbrooke, Canada, 1983.

57. Malhotra, V. M., Ed., Proc. CANMET/ACI 1st Int. Conf. on the Use of Fly Ash, Silica Fume, Slag and Other Mineral By-Products in Concrete, Vol 1 and 2, American Concrete Institute, Detroit, 1983.

58. Mather, B., Cement users' expectations with regard to blended cements, ACI Spec. Publ. SP79, Vol. 1, Malhotra, V. M., Ed., American Concrete Institute, Detroit, 1983, 255.

59. Mehta, P. K., Pozzolanic and cementitious by-products as mineral admixtures for concrete — a critical review, ACI Spec. Publ. SP79, Vol. 1, Malhotra, V. M., Ed., American Concrete Institute, Detroit, 1983, 1.

60. Peterson, O., Pop-out formation: a laboratory study with a Swedish opaline gravel, Proc. 6th Int. Conf. on Alkalies in Concrete, Copenhagen, Danish Concrete Association, 1983, 291.
61. Rath, G., The operation and design of electric reduction furnaces for the production of ferrosilicon and silicon metal, in *Condensed Silica Fume,* Aitcin, P. C., Ed., University of Sherbrooke, Sherbrooke, Canada, 1983.
62. Rau, G. and Aitcin, P. C., Different types of condensed silica fume, in *Condensed Silica Fume,* Aitcin, P. C., Ed., University of Sherbrooke, Sherbrooke, Canada, 1983, 9.
63. Warris, B., Strength of concrete containing secondary cementing materials, ACI Spec. Publ. SP79, Vol. 1, Malhotra, V. M., Ed., American Concrete Institute, Detroit, 1983, 539.
64. Concrete: micro-silica adds to the mix, *Civil Eng. (London),* August, 43, 1983.
65. High-strength shotcrete, *Eng. News-Rec.,* September 1983.
66. Microsilica shotcrete: new in concrete technology, *Civil Eng.,* 53(10), 1983.
67. Problematic waste product gives concrete new properties, *Concrete Prod.,* 86(10), 34, 1983.
68. Gjorv, O. E., New developments in concrete technology and their implications for structures of the future, Conference Proceedings, New Directions in Construction, University of California, Berkeley, 1984, 25.
69. Hill, T. B., Underwater repair and protection of offshore structures, *Concrete (London),* 18(5), 16, 1984.
70. Hyland, E. J., Radjy, F. F., and Sorensen, E. V., Microsilica and chemical admixtures in concrete, *Dodge Constr. News,* 30(59), 3, 1984.
71. Krantz, G. W., Selected Pneumatic Gunites for Use in Underground Mining: A Comparative Engineering Analysis, Bureau of Mines Information Circular, Paper No. 8984, U.S. Bureau of Mines, Washington, D.C., 1984.
72. Malhotra, V. M., Use of mineral admixtures for specialized concretes, *Concrete Int. Design Constr.,* 6(4), 19, 1984.
73. Philleo, R. E., Versatility and high strength in concrete materials, *Concrete Int. Design Constr.,* 6(2), 41, 1984.
74. Satkowski, J. A., Scheetz, B., and Rizer, J. M., Cementitious Composite Material With Stainless Steel Particulate Filler, U.S. Patent 4,482,385, 1984.
75. Saucier, K. L., High-Strength Concrete for Peacekeeper Facilities, WES Misc. Paper SL-84-3, U.S. Army Corps of Engineers Waterways Experiment Station, Vicksburg, Miss., 1984.
76. Aitcin, P. C., Ballivy, G., and Parizeau, R., The Use of Condensed Silica Fume in Grouts, Innovative Cement Grouting, ACI Spec. Publ. 83, American Concrete Institute, Detroit, 1985, 1.
77. Krenchel, H. and Shah, S., Applications of polypropylene fibers in Scandinavia, *Concrete Int. Design Constr.,* 7(3), 32, 1985.
78. Ohama, Y., Amano, M., and Endo, M., Properties of carbon fiber reinforced cement with silica fume, *Concrete Int. Design Constr.,* 7(3), 58, 1985.

INDEX